Electroacoustic Devices:
Microphones and Loudspeakers

Electroacoustic Devices: Microphones and Loudspeakers

Edited by
Glen Ballou

Routledge
Taylor & Francis Group

LONDON AND NEW YORK

First published 2009
This edition published 2013 by Focal Press

Published 2017 by Routledge
2 Park Square, Milton Park, Abingdon, Oxon OX14 4RN
711 Third Avenue, New York, NY 10017, USA

First issued in hardback 2017

Routledge is an imprint of the Taylor and Francis Group, an informa business

Notices
Practitioners and researchers must always rely on their own experience and knowledge in evaluating and using any information, methods, compounds, or experiments described herein. In using such information or methods they should be mindful of their own safety and the safety of others, including parties for whom they have a professional responsibility.

Product or corporate names may be trademarks or registered trademarks, and are used only for identification and explanation without intent to infringe.

Library of Congress Cataloging-in-Publication Data
Application submitted

Typeset by: diacriTech, Chennai, India

ISBN 13: 978-1-138-40658-2 (hbk)
ISBN 13: 978-0-240-81267-0 (pbk)

Contents

Part I
Electroacoustic Devices

1. ### Microphones
   Glen Ballou, Joe Ciaudelli, Volker Schmitt

2. Loudspeakers
Jay Mitchell

3. Loudspeaker Cluster Design
Ralph Heinz

Part I

Electroacoustic Devices

Microphones

Glen Ballou, Joe Ciaudelli, Volker Schmitt

1.1 INTRODUCTION

All sound sources have different characteristics; their waveform varies, their phase characteristics vary, their dynamic range and attack time vary, and their frequency response varies, just to name a few. No one microphone will reproduce all of these characteristics equally well. In fact, each sound source will sound better or more natural with one type or brand of microphone than all others. For this reason we have and always will have many types and brands of microphones.

Microphones are electroacoustic devices that convert acoustical energy into electrical energy. All microphones have a diaphragm or moving surface that is excited by the acoustical wave. The corresponding output is an electrical signal that represents the acoustical input.

Microphones fall into two classes: *pressure* and *velocity*. In a *pressure* microphone the diaphragm has only one surface exposed to the sound source so the output corresponds to the instantaneous sound pressure of the impressed sound waves. A pressure microphone is a *zero-order gradient microphone*, and is associated with omnidirectional characteristics.

The second class of microphone is the *velocity* microphone, also called a *first-order gradient microphone*, where the effect of the sound wave is the difference or gradient between the sound wave that hits the front and the rear of the diaphragm. The electrical output corresponds to the instantaneous particle velocity in the impressed sound wave. Ribbon microphones as well as pressure microphones that are altered to produce front-to-back discrimination are of the velocity type.

Microphones are also classified by their pickup pattern or how they discriminate between the various directions the sound source comes from, Fig. 1.1. These classifications are:

- *Omnidirectional*—pickup is equal in all directions.
- *Bidirectional*—pickup is equal from the two opposite directions (180°) apart and zero from the two directions that are 90° from the first.
- *Unidirectional*—pickup is from one direction only, the pickup appearing cardioid or heart-shaped.

The air particle relationships of the air particle displacement, velocity, and acceleration that a microphone sees as a plane wave in the far field, are shown in Fig. 1.2.

1.2 PICKUP PATTERNS

Microphones are made with single- or multiple-pickup patterns and are named by the pickup pattern they employ. The pickup patterns and directional response characteristics of the various types of microphones are shown in Fig. 1.1.

1.2.1 Omnidirectional Microphones

The omnidirectional, or spherical, polar response of the pressure microphones, Fig. 1.3, is created because the diaphragm is only exposed to the acoustic wave on the front side. Therefore, no cancellations are produced by having sound waves hitting both the front and rear of the diaphragm at the same time.

Omnidirectional microphones become increasingly directional as the diameter of the microphone reaches the wavelength of the frequency in question, as shown in Fig. 1.4; therefore, the microphone should have the smallest diameter

Microphone	Omnidirectional	Bidirectional	Directional	Supercardioid	Hypercardioid
Directional Response Characteristics	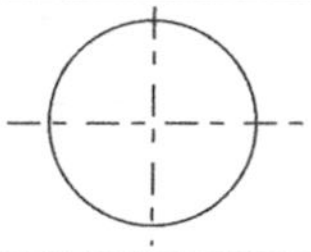	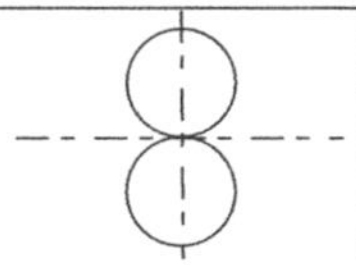	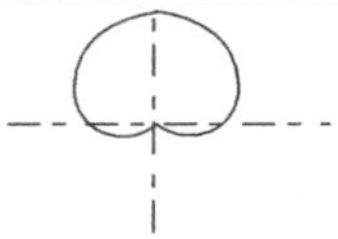	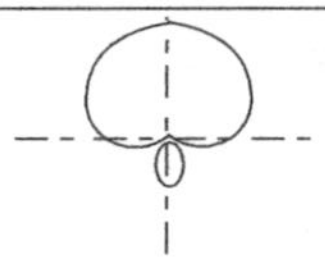	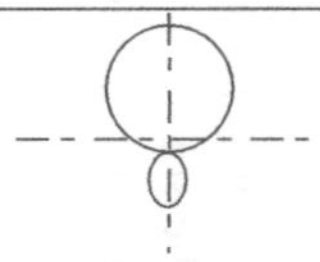
Voltage output	$E = E_0$	$E = E_0 \cos\theta$	$E = \frac{E_0}{2}(1 + \cos\theta)$	$E = \frac{E_0}{2}[(\sqrt{3}-1) + (3-\sqrt{3})\cos\theta]$	$E = \frac{E_0}{4}(1 + 3\cos\theta)$
Random energy efficiency (%)	100	33	33	27	25
Front response / Back response	1	1	∞	3.8	2
Front random response / Total random response	0.5	0.5	0.67	0.93	0.87
Front random response / Back random response	1	1	7	14	7
Equivalent distance	1	1.7	1.7	1.9	2
Pickup angle (2θ) for 3 dB attenuation	—	90°	130°	116°	100°
Pickup angle (2θ) for 6 dB attenuation	—	120°	180°	156°	140°

FIGURE 1.1 Performance characteristics of various microphones.

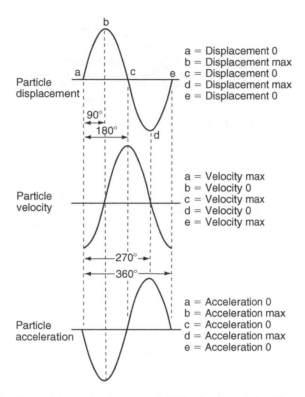

FIGURE 1.2   Air particle motion in a sound field, showing relationship to velocity and acceleration.

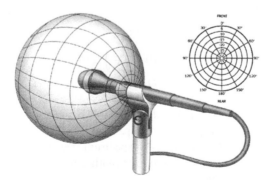

FIGURE 1.3   Omnidirectional pickup pattern. Courtesy Shure Incorporated.

possible if omnidirectional characteristics are required at high frequencies. The characteristic that allows waves to bend around objects is known as *diffraction* and happens when the wavelength is long compared to the size of the object.

PART | I   Electroacoustic Devices

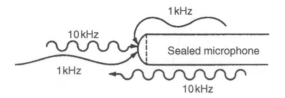

FIGURE 1.4   High-frequency directivity of an omnidirectional microphone.

As the wavelength approaches the size of the object, the wave cannot bend sharply enough and, therefore, passes by the object. The various responses start to diverge at the frequency at which the diameter of the diaphragm of the microphone, D, is approximately one-tenth the wavelength, λ, of the sound arriving as equation

$$D = \frac{\lambda}{10} \qquad (1.1)$$

The frequency, f, at which the variation begins is

$$f = \frac{v}{10D} \qquad (1.2)$$

where,
v is the velocity of sound in feet per second, or meters per second,
D is the diameter of the diaphragm in feet or meters.

For example, a ½ inch (1.27 cm) microphone will begin to vary from omnidirectional, though only slightly, at

$$f = \frac{1130}{\left(10 \times \dfrac{0.5}{12}\right)}$$
$$= 2712\,\mathrm{Hz}$$

and will be down approximately 3 dB at 10,000 Hz.

Omnidirectional microphones are capable of having a very flat, smooth frequency response over the entire audio spectrum because only the front of the diaphragm is exposed to the source, eliminating phase cancellations found in unidirectional microphones.

For smoothness of response the smaller they are, the better. The problem usually revolves around the smallest diaphragm possible versus the lowest signal-to-noise ratio, SNR, or put another way, the smaller the diaphragm, the lower the microphone sensitivity, therefore, the poorer the SNR.

Omnidirectional microphones have very little proximity effect. See Section 1.2.3.1 for a discussion on proximity effect.

Because the pickup pattern is spherical, the random energy efficiency is 100%, and the ratio of front response to back or side is 1:1; therefore, signals from the sides or rear will have the same pickup sensitivity as from the front, giving a directivity index of 0dB. This can be helpful in picking up wanted room characteristics or conversations around a table as when recording a symphony. However, it can be detrimental when in a noisy environment.

Omnidirectional microphones are relatively free from mechanical shock because the output at all frequencies is high; therefore, the diaphragm can be stiff. This allows the diaphragm to follow the magnet or stationary system it operates against when subjected to mechanical motion (see Section 1.3.3).

1.2.2 Bidirectional Microphones

A *bidirectional microphone* is one that picks up from the front and back equally well with little or no pickup from the sides. The field pattern, Fig. 1.5, is called a *figure 8*.

Because the microphone discriminates between the front, back, and sides, random energy efficiency is 33%. In other words, background noise, if it is in a reverberant field, will be 67% lower than with an omnidirectional microphone. The front-to-back response will still remain one; however, the front-to-side response will approach infinity, producing a directivity index of 4.8. This can be extremely useful when picking up two conversations on opposite sides of a table. Because of the increased directional capabilities of the microphone, pickup distance is 1.7 times greater before feedback in the direct field than for an omnidirectional microphone. The included pickup cone angle shown in Fig. 1.6 for 6dB attenuation on a perfect bidirectional microphone is 120° off the front of the microphone and 120° off the rear of the microphone. Because of diffraction, this angle varies with frequency, becoming narrower as the frequency increases.

FIGURE 1.5   Bidirectional pickup pattern. Courtesy Sennheiser Electronic Corporation.

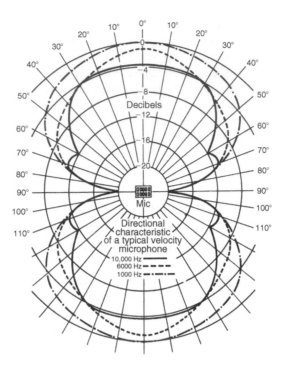

FIGURE 1.6   Polar pattern of a typical bidirectional ribbon velocity microphone showing the narrowing pattern at high frequencies.

1.2.3 Unidirectional Microphones

Unidirectional microphones have a greater sensitivity to sound pickup from the front than any other direction.

The average unidirectional microphone has a front-to-back ratio of 20–30 dB; that is, it has 20–30 dB greater sensitivity to sound waves approaching from the front than from the rear.

Unidirectional microphones are usually listed as *cardioid* or *directional*, Fig. 1.7, *supercardioid*, Fig. 1.8, or *hypercardioid*, Fig. 1.9. The pickup pattern is called *cardioid* because it is heart shaped. Unidirectional microphones are the most commonly used microphones because they discriminate between signal and random unwanted noise. This has many advantages including:

- Less background noise,
- More gain before feedback especially when used in the direct field,
- Discrimination between sound sources.

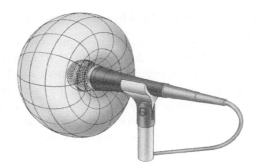

FIGURE 1.7   Cardioid pickup pattern. Courtesy Shure Incorporated.

FIGURE 1.8   Supercardioid pickup pattern. Courtesy Shure Incorporated.

FIGURE 1.9   Hypercardioid pickup pattern. Courtesy Sennheiser Electronic Corporation.

The cardioid pattern can be produced by one of two methods:

1. The first method combines the output of a pressure diaphragm and a pressure-gradient diaphragm, as shown in Fig. 1.10. Since the pressure-gradient diaphragm has a bidirectional pickup pattern and the pressure diaphragm has an omnidirectional pickup pattern, the wave hitting the front of the diaphragms adds, while the wave hitting the rear of the diaphragm cancels as it is 180° out-of-phase with the rear pickup pattern of the pressure diaphragm. This method is expensive and seldom used for sound reinforcement or general-purpose microphones.

2. The second and most widely used method of producing a cardioid pattern is to use a single diaphragm and acoustically delay the wave reaching the rear of the diaphragm. When a wave approaches from the front of the diaphragm, it first hits the front and then the rear of the diaphragm after traveling through an acoustical delay circuit, as shown in Fig. 1.11A. The pressure on the front of the diaphragm is at 0° while on the rear of the diaphragm it is some angle between 0° and 180°, as shown in Fig. 1.11B. If the rear pressure was at 0°, the output would be 0. It would be ideal if the rear pressure were at 180° so that it could add to the input, doubling the output.

The phase inversion is caused by the extra distance the wave has to travel to reach the back of the diaphragm. When the wave is coming from the rear of

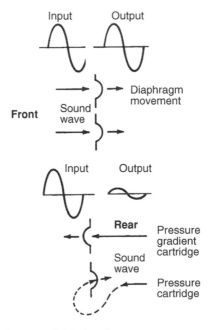

FIGURE 1.10   Two-diaphragm cardioid microphone.

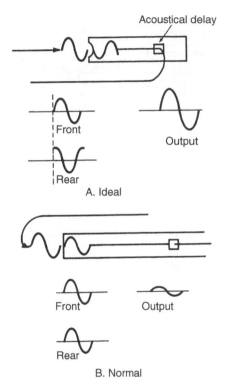

FIGURE 1.11 Cardioid microphone employing acoustical delay.

the microphone, it hits the front and back of the diaphragm at the same time and with the same polarity, therefore canceling the output.

The frequency response of cardioid microphones is usually rougher than an omnidirectional microphone due to the acoustical impedance path and its effects on the front wave response. The front and rear responses of a cardioid microphone are not the same. Although the front pattern may be essentially flat over the audio spectrum, the back response usually increases at low and high frequencies, as shown in Fig. 1.12.

Discrimination between the front and back response is between 15 and 30 dB in the mid frequencies and could be as little as 5–10 dB at the extreme ends, as shown in Fig. 1.12.

1.2.3.1 Proximity Effects

As the source is moved closer to the diaphragm, the low-frequency response increases due to the proximity effect, Fig. 1.13. The proximity effect[1] is created because at close source-to-microphone distance, the magnitude of the

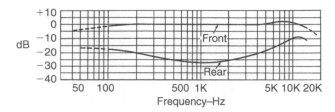

FIGURE 1.12   Frequency response of a typical cardioid microphone.

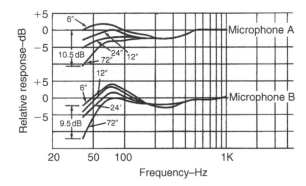

FIGURE 1.13   Proximity effect variations in response to distance between source and microphone for cardioid microphones. Courtesy Telex Electro-Voice.

sound pressure on the front is appreciably greater than the sound pressure on the rear. In the vector diagram shown in Fig. 1.14A, the sound source was a distance greater than 2 ft from the microphone. The angle $2KD$ is found from D, which is the acoustic distance from the front to the rear of the diaphragm and $K = 2\pi/\lambda$. Figure 1.14B shows the vector diagram when used less than 4 inches to the sound source.

In both cases, force F_1, the sound pressure on the front of each diaphragm, is the same. Force F_2 is the force on the back of the diaphragm when the microphone is used at a distance from the sound source, and F_0 is the resultant force. The force F_0' on the back of the diaphragm is created by a close sound source. Laterally, the vector sum F_0' is considerably larger in magnitude than F_0 and therefore produces greater output from the microphone at low frequencies. This can be advantageous or disadvantageous. It is particularly useful when vocalists want to add low-frequency to their voice or an instrumentalist to add low frequencies to the instrument. This is accomplished by varying the distance between the microphone and the sound source, increasing bass as the distance decreases.

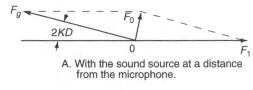

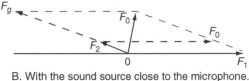

FIGURE 1.14   Vector diagram of a unidirectional microphone. Courtesy Telex Electro-Voice.

1.2.3.1.1 Frequency Response

Frequency response is an important specification of unidirectional microphones and must be carefully analyzed and interpreted in terms of the way the microphone is to be used. If a judgment as to the sound quality of the microphone is made strictly from a single on-axis response, the influence of the proximity effect and off-axis response would probably be overlooked. A comparison of frequency response as a function of microphone-to-source distance will reveal that *all* unidirectional microphones experience a certain amount of proximity effect. In order to evaluate a microphone, this variation with distance is quite important.

When using a unidirectional microphone[2] in a handheld or stand-mounted configuration, it is conceivable that the performer will not always remain exactly on axis. Variations of ±45° often occur, and so a knowledge of the uniformity of response over such a range is important. The nature of these response variations is shown in Fig. 1.15. Response curves such as these give a better indication of this type of off-axis performance than polar response curves. The polar response curves are limited in that they are usually given for only a few frequencies, therefore, the complete spectrum is difficult to visualize.

For applications involving feedback control or noise rejection, the polar response or particular off-axis response curves, such as at 135° or 180°, are important. These curves can often be misleading due to the acoustic conditions and excitation signals used. Such measurements are usually made under anechoic conditions at various distances with sine-wave excitation. Looking solely at a rear response curve as a function of frequency is misleading since such a curve does not indicate the polar characteristic at any particular frequency, but only the level at one angle. Such curves also tend to give the impression of a rapidly fluctuating high-frequency discrimination. This sort of performance is to be expected since it is virtually impossible to design a microphone of practical size with a constant angle of best discrimination at high frequencies, Fig. 1.16. The principal factor influencing this variation in rear response is diffraction, which

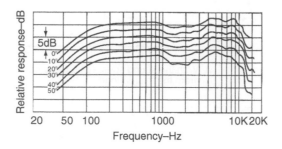

FIGURE 1.15 Variations in front response versus angular position. Note: Curves have been displaced by 2.5 dB for comparison purposes.

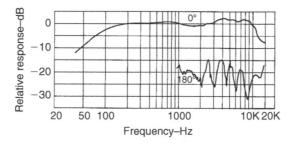

FIGURE 1.16 Typical fluctuations in high-frequency rear response for a cardioid microphone. Courtesy Shure Incorporated.

is caused by the physical presence of the microphone in the sound field. This diffraction effect is frequency dependent and tends to disrupt the ideal performance of the unidirectional phase-shift elements.

To properly represent this high-frequency off-axis performance, a polar response curve is of value, but it, too, can be confusing at high frequencies. The reason for this confusion can be seen in Fig. 1.17, where two polar response curves only 20 Hz apart are shown. The question that arises then is how can such performance be properly analyzed? A possible solution is to run polar response curves with bands of random noise such as ⅓ octaves of pink noise. Random noise is useful because of its averaging ability and because its amplitude distribution closely resembles program material.

Anechoic measurements are only meaningful as long as no large objects are in close proximity to the microphone. The presence of the human head in front of a microphone will seriously degrade the effective high-frequency discrimination. An example of such degradation can be seen in Fig. 1.18 where a head object was placed 2 inches (5 cm) in front of the microphone. (The two curves have not been normalized.) This sort of performance results from the head as a reflector and is a common cause of feedback as one approaches a microphone. This should not be considered as a shortcoming of the microphone, but rather

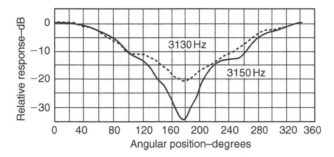

FIGURE 1.17   An example of rapid variations in high-frequency polar response for single-frequency excitation. Courtesy Shure Incorporated.

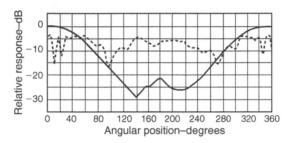

FIGURE 1.18   An example of a head obstacle on a polar response. Courtesy Shure Incorporated.

as an unavoidable result of the sound field in which it is being used. At 180°, for example, the microphone sees, in addition to the source it is trying to reject, a reflection of that source some 2 inches (5 cm) in front of its diaphragm. This phenomenon is greatly reduced at low frequencies because the head is no longer an appreciable obstacle to the sound field. It is thus clear that the effective discrimination of any unidirectional microphone is greatly influenced by the sound field in which it is used.

1.2.3.1.2 Types of Cardioid Microphones

Cardioid microphones are named by the way sound enters the rear cavity. The sound normally enters the rear of the microphone's cavity through single or multiple holes in the microphone housing, as shown in Fig. 1.19.

1.2.3.1.3 Single-Entry Cardioid Microphones

All single-entrant cardioid microphones have the rear entrance port located at one distance from the rear of the diaphragm. The port location is usually within 1 ½ inches (3.8 cm) of the diaphragm and can cause a large proximity effect. The Electro-Voice DS35 is an example of a single-entrant cardioid microphone, Fig. 1.20.

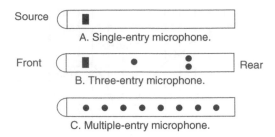

A. Single-entry microphone.

B. Three-entry microphone.

C. Multiple-entry microphone.

FIGURE 1.19   Three types of cardioid microphones.

FIGURE 1.20   Electro-Voice DS35 single-entrant microphone. Courtesy Electro-Voice, Inc.

The low-frequency response of the DS35 varies as the distance from the sound source to the microphone decreases, Fig. 1.21. Maximum bass response is produced in close-up use with the microphone 1 ½ inches (3.8 cm) from the sound source. Minimum bass response is experienced at distances greater than 2 ft (0.6 m). Useful effects can be created by imaginative application of the variable low-frequency response.

Another single-entrant microphone is the Shure SM81.[3] The acoustical system of the microphone operates as a first-order gradient microphone with two sound openings, as shown in Fig. 1.22. Figure 1.22 shows a simplified cross-sectional view of the transducer, and Fig. 1.23 indicates the corresponding electrical analog circuit of the transducer and preamplifier.

One sound opening, which is exposed to the sound pressure p_1, is represented by the front surface of the diaphragm. The other sound opening, or rear entry, consists of a number of windows in the side of the transducer housing where the sound pressure p_2 prevails. The diaphragm has an acoustical impedance Z_0, which also includes the impedance of the thin air film between the diaphragm and backplate. The sound pressure p_2 exerts its influence on the rear surface

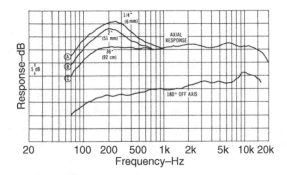

FIGURE 1.21   Frequency response versus distance for an Electro-Voice DS35 single-entrant cardioid microphone. Courtesy Electro-Voice, Inc.

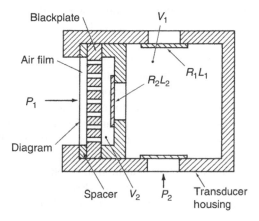

FIGURE 1.22   Simplified cross-sectional view of the Shure SM81 condenser transducer. Courtesy Shure Incorporated.

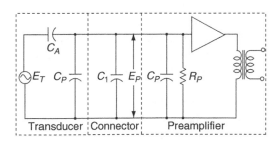

FIGURE 1.23   Electrical equivalent circuit of the Shure SM81 condenser transducer and preamplifier. Courtesy Shure Incorporated.

of the diaphragm via a screen mounted in the side windows of the transducer housing, having a resistance R_1 and inertance L_1, through the cavity V_1 with compliance C_1. A second screen has a resistance R_2 and inertance L_2, through a second cavity V_2 with compliance C_2, and finally through the perforations in the backplate.

The combination of circuit elements L_1, R_1, C_1, L_2, R_2, C_2 forms a ladder network with lossy inertances, and is called a *lossy ladder network*. The transfer characteristic of this network enforces a time delay on the pressure p_2 imparting directional (cardioid) characteristics for low and medium frequencies. At high frequencies the attenuation caused by the network is large, and the resulting pressure arriving at the back of the diaphragm due to p_2 is small. The microphone then operates much like an omnidirectional system under the predominant influence of p_1. At these frequencies directional characteristics are attained by diffraction of the sound around a suitably shaped transducer housing.

A rotary low-frequency response shaping switch allows the user to select between a flat and a 6 dB/octave roll-off at 100 Hz or an 18 dB/octave cutoff at 80 Hz. The 100 Hz roll-off compensates for the proximity effect associated with a 6 inches (15 cm) source to microphone distance, while the 80 Hz cutoff significantly reduces most low-frequency disturbances with minimal effect on voice material. In the flat position the microphone has a 6 dB/octave electronic infrasonic roll-off, with −3 dB at 10 Hz to reduce the effects of inaudible low-frequency disturbances on microphone preamplifier inputs. Attenuation is provided for operation at high sound pressure levels (to 145 dB SPL) by means of a rotary capacitive switch (see Section 1.3.4.1).

A final example of a single entry cardioid microphone is the Shure Beta 57 supercardioid dynamic microphone, Fig. 1.24. Both the Shure Beta 57 and the Beta 58 use neodymium magnets for hotter output and incorporate an improved shock mount.

1.2.3.1.4 Three-Entry Cardioid Microphones

The Sennheiser MD441 is an example of a three-entry cardioid microphone, Fig. 1.25. The low-frequency rear entry has a d (distance from center of diaphragm to the entry port) of about 2.8 inches (7 cm), the mid-frequency entry d is about 2.2 inches (5.6 cm), and the high-frequency entry d is about 1.5 inches (3.8 cm), with the transition in frequency occurring between 800 Hz and 1 kHz. Each entry consists of several holes around the microphone case rather than a single hole.

This configuration is used for three reasons. By using a multiple arrangement of entry holes around the circumference of the microphone case into the low-frequency system, optimum front response and polar performance can be maintained, even though most of the entries may be accidentally covered when the microphone is handheld or stand mounted. The microphone has good proximity performance because the low-frequency entry ports are far from

FIGURE 1.24   Shure Beta 57 dynamic microphone. Courtesy Shure Incorporated.

FIGURE 1.25   Sennheiser MD441 three-entry cardioid microphone. Courtesy Sennheiser Electronic Corporation.

the diaphragm (4.75 in) as well as the high-frequency entry having very little proximity influence at low frequencies. The two-entry configuration has a cardioid polar response pattern that provides a wide front working angle as well as excellent noise rejection and feedback control.

1.2.3.1.5 Multiple-Entry Cardioid Microphones

The Electro-Voice RE20 Continuously Variable-D microphone, Fig. 1.26, is an example of multiple-entry microphones. Multiple-entry microphones have many rear entrance ports. They can be constructed as single ports, all at a different distance from the diaphragm, or as a single continuous opening port. Each entrance is tuned to a different band of frequencies, the port closest to the diaphragm being tuned to the high frequencies, and the port farthest from the diaphragm being tuned to the low-frequency band. The greatest advantage of this arrangement is reduced-proximity effect because of the large distance between the source and the rear entry low-frequency port, and mechanical crossovers are not as sharp and can be more precise for the frequencies in question.

As in many cardioid microphones, the RE20 has a low-frequency roll-off switch to reduce the proximity effect when close micing. Figure 1.27 shows the wiring diagram of the RE20. By moving the red wire to either the $150\,\Omega$ or $50\,\Omega$ tap, the microphone output impedance can be changed. Note the "bass tilt" switch that, when open, reduces the series inductance and, therefore, the low-frequency response.

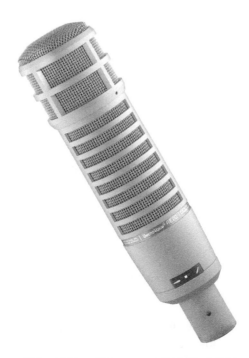

FIGURE 1.26 Electro-Voice RE20 multiple-entry (variable-D cardioid microphone). Courtesy Telex Electro-Voice.

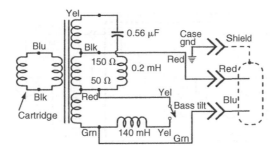

FIGURE 1.27   Electro-Voice RE20 cardioid wiring diagram. Note "bass tilt" switch circuit and output impedance taps. Courtesy Telex Electro-Voice.

1.2.3.1.6 Two-Way Cardioid Microphones

In a two-way microphone system, the total response range is divided between a high-frequency and a low-frequency transducer, each of which is optimally adjusted to its specific range similar to a two-way loudspeaker system. The two systems are connected by means of a crossover network.

The AKG D-222EB schematically shown in Fig. 1.28 employs two coaxially mounted dynamic transducers. One is designed for high frequencies and is placed closest to the front grille and facing forward. The other is designed for low frequencies and is placed behind the first and facing rearward. The low-frequency transducer incorporates a hum-bucking winding to cancel the effects of stray magnetic fields. Both transducers are coupled to a 500 Hz inductive-capacitive-resistive crossover network that is electroacoustically phase corrected and factory preset for linear off-axis response. (This is essentially the same design technique used in a modern two-way loud-speaker system.)

The two-way microphone has a predominantly frequency-independent directional pattern, producing more linear frequency response at the sides of the microphone and far more constant discrimination at the rear of the microphone. Proximity effect at working distances down to 6 inches (15 cm) is reduced because the distance between the microphone windscreen and the low-frequency transducer is large.

The D-222EB incorporates a three-position bass-roll-off switch that provides 6 dB or 12 dB attenuation at 50 Hz. This feature is especially useful in speech applications and in acoustically unfavorable environments with excessive low-frequency ambient noise, reverberation, or feedback.

1.3 TYPES OF TRANSDUCERS

1.3.1 Carbon Microphones

One of the earliest types of microphones, the *carbon microphone*, is still found in old telephone handsets. It has very limited frequency response, is very noisy, has high distortion, and requires a hefty dc power supply. A carbon microphone[4] is shown in Fig. 1.29 and operates in the following manner.

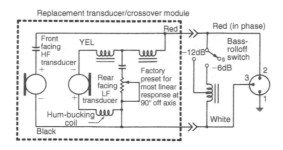

FIGURE 1.28   Schematic of an AKG D-222EB two-way cardioid microphone. Courtesy AKG Acoustics, Inc.

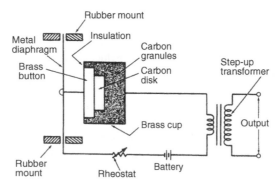

FIGURE 1.29   Connection and construction of a single-button carbon microphone.

Several hundred small carbon granules are held in close contact in a brass cup called a *button* that is attached to the center of a metallic diaphragm. Sound waves striking the surface of the diaphragm disturb the carbon granules, changing the contact resistance between their surfaces. A battery or dc power source is connected in series with the carbon button and the primary of an audio impedance-matching transformer. The change in contact resistance causes the current from the power supply to vary in amplitude, resulting in a current waveform similar to the acoustic waveform striking the diaphragm.

The impedance of the carbon button is low so a step-up transformer is used to increase the impedance and voltage output of the microphone and to eliminate dc from the output circuit.

1.3.2 Crystal and Ceramic Microphones

Crystal and *ceramic* microphones were once popular because they were inexpensive and their high-impedance high-level output allowed them to be connected directly to the input grid of a tube amplifier. They were most popular in use with home tape recorders where microphone cables were short and input impedances high.

Crystal and ceramic microphones[5] operate as follows: piezoelectricity is "pressure electricity" and is a property of certain crystals such as Rochelle salt, tourmaline, barium titanate, and quartz. When pressure is applied to these crystals, electricity is generated. Present-day commercial materials such as ammonium dihydrogen phosphate (ADP), lithium sulfate (LN), dipotassium tartrate (DKT), potassium dihydrogen phosphate (KDP), lead zirconate, and lead titanate (PZT) have been developed for their piezoelectric qualities. Ceramics do not have piezoelectric characteristics in their original state, but the characteristics are introduced in the materials by a polarizing process. In piezoelectric ceramic materials the direction of the electrical and mechanical axes depends on the direction of the original dc polarizing potential. During polarization a ceramic element experiences a permanent increase in dimensions between the poling electrodes and a permanent decrease in dimension parallel to the electrodes.

The crystal element can be cut as a bender element that is only affected by a bending motion or as a twister element that is only affected by a twisting motion, Fig. 1.30.

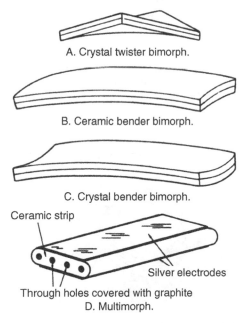

FIGURE 1.30   Curvatures of bimorphs and multimorph. Courtesy Clevite Corp., Piezoelectric Division.

The internal capacitance of a crystal microphone is about $0.03\,\mu F$ for the diaphragm-actuated type and 0.0005–$0.015\,\mu F$ for the sound-cell type.

The *ceramic* microphone operates like a crystal microphone except it employs a barium titanate slab in the form of a ceramic, giving it better temperature and humidity characteristics.

Crystal and ceramic microphones normally have a frequency response from 80 to 6500 Hz but can be made to have a flat response to 16 kHz. Their output impedance is about $100\,k\Omega$, and they require a minimum load of 1–$5\,M\Omega$ to produce a level of about -30 dB re 1 V/Pa.

1.3.3 Dynamic Microphones

The *dynamic microphone* is also referred to as a *pressure* or *moving-coil* microphone. It employs a small diaphragm and a voice coil, moving in a permanent magnetic field. Sound waves striking the surface of the diaphragm cause the coil to move in the magnetic field, generating a voltage proportional to the sound pressure at the surface of the diaphragm.

In a dynamic pressure unit, Fig. 1.31, the magnet and its associated parts (magnetic return, pole piece, and pole plate) produce a concentrated magnetic flux of approximately 10,000 G in the small gap.

The diaphragm, a key item in the performance of a microphone, supports the voice coil centrally in the magnetic gap, with only 0.006 inch clearance.

An omnidirectional diaphragm and voice-coil assembly is shown in Fig. 1.32. The compliance section has two hinge points with the section between them made up of tangential corrugated triangular sections that stiffen this portion and allow the diaphragm to move in and out with a slight rotating motion. The hinge points are designed to permit high-compliance action. A spacer supports the moving part of the diaphragm away from the top pole plate to provide room for its movement. The cementing flat is bonded to the face plate. A stiff

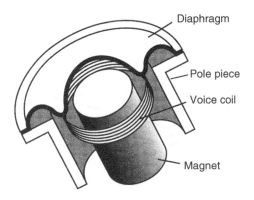

FIGURE 1.31   A simplified drawing of a dynamic microphone. Courtesy Shure Incorporated.

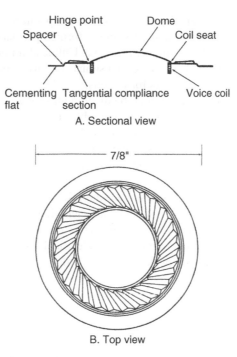

FIGURE 1.32   Omnidirectional diaphragm and voice coil assembly.

hemispherical dome is designed to provide adequate acoustical capacitance. The coil seat is a small step where the voice coil is mounted, centered, and bonded on the diaphragm.

Early microphones had aluminum diaphragms that were less than 1 mil (0.001 in) thick. Aluminum is light-weight, easy to form, maintains its dimensional stability, and is unaffected by extremes in temperature or humidity. Unfortunately, being only 1 mil thick makes the diaphragms fragile. When it is touched or otherwise deformed by excessive pressure, an aluminum diaphragm is dead.

Mylar™, a polyester film manufactured by the DuPont Company, is commonly used for diaphragms. Mylar™ is a unique plastic. Extremely tough, it has high tensile strength, high resistance to wear, and outstanding flex life. Mylar™ diaphragms have been cycle tested with temperature variations from –40°F to +170°F (–40°C to +77°C) over long periods without any impairment to the diaphragm. Since Mylar™ is extremely stable, its properties do not change within the temperature and humidity range in which microphones are used.

The specific gravity of Mylar™ is approximately 1.3 as compared to 2.7 for aluminum so a Mylar™ diaphragm may be made considerably thicker without upsetting the relationship of the diaphragm mass to the voice-coil mass.

Mylar™ diaphragms are formed under high temperature and high pressure, a process in which the molecular structure is formed permanently to establish a *dimensional memory* that is highly retentive. Unlike aluminum, Mylar™ diaphragms will retain their shape and dimensional stability although they may be subjected to drastic momentary deformations.

The voice coil weighs more than the diaphragm so it is the controlling part of the mass in the diaphragm voice-coil assembly. The voice coil and diaphragm mass (analogous to inductance in an electrical circuit) and compliance (analogous to capacitance), make the assembly resonate at a given frequency as any tuned electrical circuit. The free-cone resonance of a typical undamped unit is in the region of 350 Hz.

If the voice coil were left undamped, the response of the assembly would peak at 350 Hz, Fig. 1.33. The resonant characteristic is damped out by the use of an acoustic resistor, a felt ring that covers the openings in the centering ring behind the diaphragm. This is analogous to electrical resistance in a tuned circuit. While this reduces the peak at 350 Hz, it does not fix the droop below 200 Hz. Additional acoustical resonant devices are used inside the microphone case to correct the drooping. A cavity behind the unit (analogous to capacitance) helps resonate at the low frequencies with the mass (inductance) of the diaphragm and voice-coil assembly.

Another tuned resonant circuit is added to extend the response down to 35 Hz. This circuit, tuned to about 50 Hz, is often a tube that couples the inside cavity of the microphone housing to the outside, Fig. 1.34.

The curvature of the diaphragm dome provides stiffness, and the air cavity between it and the dome of the pole piece form an acoustic capacitance. This capacitance resonates with the mass (inductance) of the assembly to extend the response up to 20 kHz.

To control the high-frequency resonance, a nonmagnetic filter is placed in front of the diaphragm, creating an acoustic resistance, Fig. 1.34. The filter is also an effective mechanical protection device. The filter prevents dirt particles, magnetic chips, and moisture from gravitating to the inside of the unit. Magnetic chips, if allowed to reach the magnetic gap area, eventually will accumulate on top of the diaphragm and impair the frequency response. It is possible

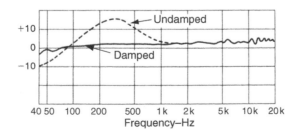

FIGURE 1.33   Diaphragm and voice-coil assembly response curve.

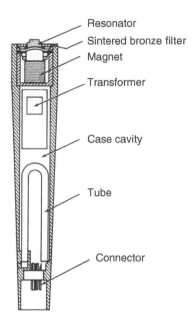

FIGURE 1.34 Omnidirectional microphone cross-section view.

for such chips to pin the diaphragm to the pole piece, rendering the microphone inoperative.

Figure 1.35 illustrates the effect of a varying sound pressure on a moving-coil microphone. For this simplified explanation, assume that a mass less diaphragm voice-coil assembly is used. The acoustic waveform, Fig. 1.35A, is one cycle of an acoustic waveform, where *a* indicates atmospheric pressure *AT*; and *b* represents atmospheric pressure plus a slight overpressure increment Δ or $AT + \Delta$.

The electrical waveform output from the moving-coil microphone, Fig. 1.35B, does not follow the phase of the acoustic waveform because at maximum pressure, $AT + \Delta$ or b, the diaphragm is at rest (no velocity). Further, the diaphragm and its attached coil reach maximum velocity, hence maximum electrical amplitude—at point *c* on the acoustic waveform. This is of no consequence unless another microphone is being used along with the moving-coil microphone where the other microphone does not see the same 90° displacement. Due to this phase displacement, condenser microphones should not be mixed with moving-coil or ribbon microphones when micing the same source at the same distance. (Sound pressure can be proportional to velocity in many practical cases.)[6]

A steady overpressure which can be considered an acoustic square wave, Fig. 1.35C, would result in the output shown in Fig. 1.35D. As the acoustic pressure rises from *a* to *b*, it has velocity, Fig. 1.35, creating a voltage output from the microphone. Once the diaphragm reaches its maximum displacement

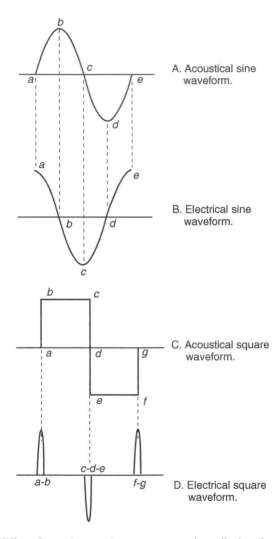

FIGURE 1.35   Effect of a varying sound pressure on a moving-coil microphone.

at *b*, and stays there during the time interval represented by the distance between *b* and *c*, voice-coil velocity is zero so electrical output voltage ceases and the output returns to zero. The same situation repeats itself from *c* to *e* and from *e* to *f* on the acoustic waveform. As can be seen, a moving-coil microphone cannot reproduce a square wave.

Another interesting theoretical consideration of the moving-coil microphone mechanism is shown in Fig. 1.36. Assume a sudden transient condition. Starting at *a* on the acoustic waveform, the normal atmospheric pressure is suddenly increased by the first wave front of a new signal and proceeds to the first

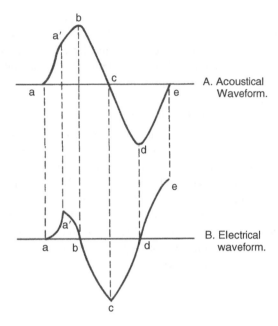

FIGURE 1.36   Effect of a transient condition on a moving-coil microphone.

overpressure peak, $AT + \Delta$ or b. The diaphragm will reach a maximum velocity halfway to b and then return to zero velocity at b. This will result in a peak, a', in the electrical waveform. From b on, the acoustic waveform and the electrical waveform will proceed as before, cycle for cycle, but 90° apart.

In this special case, peak a' does not follow the input precisely so it is something extra. It will probably be swamped out by other problems (especially mass) encountered in a practical moving-coil microphone. It does illustrate that even with a "perfect," massless, moving-coil microphone, "perfect" electrical waveforms will not be produced.

When sound waves vibrate the diaphragm, the voice coil has a voltage induced in it proportional to the magnitude and at the frequency of the vibrations. The voice coil and diaphragm have some finite mass and any mass has inertia that causes it to want to stay in the condition it is in—namely, in motion or at rest. If the stationary part of the diaphragm-magnet structure is moved in space, the inertia of the diaphragm and coil causes them to try to remain fixed in space. Therefore, there will be relative motion between the two parts with a resultant electrical output. An electrical output can be obtained in two ways, by motion of the diaphragm from airborne acoustical energy or by motion of the magnet circuit by structure-borne vibration. The diaphragm motion is the desired output, while the structure-borne vibration is undesired.

Several things may be tried to eliminate the undesired output. The mass of the diaphragm and voice coil may be reduced, but there are practical limits, or the frequency response may be limited mechanically with stiffer diaphragms or electronically with filter circuits. However, limited response makes the microphone unsuitable for broad-range applications.

1.3.3.1 Unidirectional Microphones

To reject unwanted acoustical noise such as signals emanating from the sides or rear of the microphone, unidirectional microphones are used, Fig. 1.7. Unidirectional microphones are much more sensitive to vibration relative to their acoustic sensitivity than omnidirectional types. Figure 1.37 shows a plot of vibration sensitivity versus frequency for a typical omnidirectional and unidirectional microphone with the levels normalized with respect to acoustical sensitivity.

The vibration sensitivity of the unidirectional microphone is about 15 dB higher than the omnidirectional and has a peak at about 150 Hz. The peak gives a clue to help explain the difference.

Unidirectional microphones are usually differential microphones; that is, the diaphragm responds to a pressure differential between its front and back surfaces. The oncoming sound wave is not only allowed to reach the front of a diaphragm but, through one or more openings and appropriate acoustical phase-shift networks, reaches the rear of the diaphragm. At low frequencies, the net instantaneous pressure differential causing the diaphragm to move is small compared to the absolute sound pressure, Fig. 1.38. Curve A is the pressure wave that arrives at the front of the diaphragm. Curve B is the pressure wave that reaches the rear of the diaphragm after a slight delay due to the greater distance the sound has to travel to reach the rear entry and some additional phase shift it encounters after entering. The net pressure actuating the diaphragm is curve C, which is the instantaneous difference between the two upper curves. In a typical unidirectional microphone, the differential pressure at 100 Hz will be about one-tenth of the absolute pressure or 20 dB down from the pressure an omnidirectional microphone would experience.

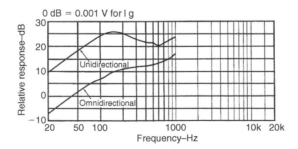

FIGURE 1.37   Vibration sensitivity of microphone cartridge.

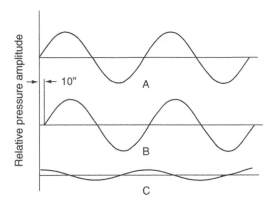

FIGURE 1.38   Differential pressure at low frequencies on unidirectional microphones.

To obtain good low-frequency response, a reasonable low-frequency electrical output is required from a unidirectional microphone. To accomplish this, the diaphragm must move more easily for a given sound pressure. Some of this is accomplished by reducing the damping resistance to less than one-tenth used in an omnidirectional microphone. This reduction in damping increases the motion of the mechanical resonant frequency of the diaphragm and voice coil, around 150 Hz in Fig. 1.37, making the microphone much more acceptable to structure-borne vibrations. Since the diaphragm of an omnidirectional microphone is much more heavily damped, it will respond less to inertial or mechanical vibration forces.

To eliminate unwanted external low-frequency noise from effecting a unidirectional microphone, some kind of isolation such as a microphone shock mount is required to prevent the microphone cartridge from experiencing mechanical shock and vibration.

1.3.4 Capacitor Microphones

In a *capacitor* or *condenser* microphone the sound pressure level varies the head capacitance of the microphone by deflecting one or two plates of the capacitor, causing an electrical signal that varies with the acoustical signal. The varying capacitance can be used to modulate an RF signal that is later demodulated or can be use d as one leg of a voltage divider, Fig. 1.39, where R and C form the voltage divider of the power supply ++ to −.

The head of most capacitor microphones consists of a small two-plate 40–50 pF capacitor. One of the two plates is a stretched diaphragm; the other is a heavy backplate or center terminal, Fig. 1.39. The backplate is insulated from the diaphragm and spaced approximately 1 mil (0.001 in) from, and parallel to, the rear surface of the diaphragm. Mathematically the output from the head may be calculated as

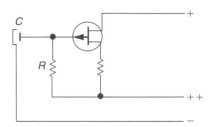

FIGURE 1.39 Voltage divider type of capacitor microphone.

$$E_O = \frac{E_p a^2 P}{8dt} \tag{1.3}$$

where,

E_O is the output in volts,

E_p is the dc polarizing voltage in volts,

a is the radius of active area of the diaphragm in centimeters,

P is the pressure in dynes per square centimeter,

d is the spacing between the backplate and diaphragm in centimeters,

t is the diaphragm tension in dynes per centimeter.

Many capacitor microphones operate with an equivalent noise level of 15–30 dB SPL. Although a 20–30 dB SPL is in the range of a well-constructed studio, a 20–30 dB microphone equivalent noise is not masked by room noise because room noise occurs primarily at low frequencies and microphone noise at high-frequency as hiss.

In the past the quality of sound recordings was limited by the characteristics of the analog tape and record material, apart from losses induced by the copying and pressing procedures. Tape saturation, for instance, created additional harmonic and disharmonic distortion components, which affected the recording fidelity at high levels, whereas the linearity at low and medium levels was quite acceptable. The onset of these distortions was rather soft and extended to a wide level range that makes it difficult to determine the threshold of audibility.

The distortion characteristics of the standard studio condenser microphones is adequate for operation with analog recording equipment. Although exhibiting a high degree of technical sophistication, these microphones show individual variations in the resolution of complex tonal structures, due to their specific frequency responses and directivity patterns and nonlinear effects inherent to these microphones.

These properties were mostly concealed by the distortions superimposed by the analog recording and playback processing. But the situation has changed essentially since the introduction of digital audio. The conversion of analog signals into digital information and vice versa is carried out very precisely, especially at high signal levels. Due to the linear quantization process the inherent distortions of digital recordings virtually decrease at increasing recording levels, which turns former distortion behavior upside down. This new reality, which is in total contrast to former experience with analog recording technique, contributes mostly to the fact that the specific distortion characteristics of the microphone may become obvious, whereas they have been masked previously by the more significant distortions of analog recording technique.

Another feature of digital audio is the enlarged dynamic range and reduced noise floor. Unfortunately, due to this improvement, the inherent noise of the microphones may become audible, because it is no longer covered up by the noise of the recording medium.

The capacitor microphone has a much faster rise time than the dynamic microphone because of the significantly lower mass of the moving parts (diaphragm versus diaphragm/coil assembly). The capacitor rise time rises from 10% of its rise time to 90% in approximately 15 μs, while the rise time for the dynamic microphone is in the order of 40 μs.

Capacitor microphones generate an output electrical waveform in step or phase with the acoustical waveform and can be adapted to measure essentially dc overpressures, Fig. 1.40.

Some advantages of capacitor microphones are:

- Small, low-mass rigid diaphragms that reduce vibration pickup.
- Smooth, extended-range frequency response.
- Rugged—capable of measuring very high sound pressure levels (rocket launches).
- Low noise (which is partially cancelled by the need for electronics).
- Small head size, which provides low diffraction interference.

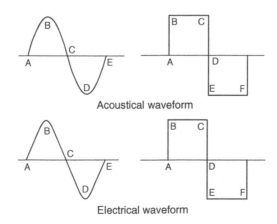

FIGURE 1.40   Capacitor microphone acoustic wave and electrical signals. Note the in-phase condition.

1.3.4.1 Voltage Divider Capacitor Microphone

Voltage divider-type capacitor microphones require a preamplifier as an integral part of the housing and a source of polarizing voltage for the head, and a source of power.

A high-quality capacitor microphone, the Sennheiser K6 Modular Condenser Microphone Series is suitable for recording studios, television and radio broadcast, motion picture studios, and stage and concert hall applications, as well as high-quality commercial sound installations.

The K6/ME62 series is a *capacitor microphone system*, Fig. 1.41, that uses AF circuitry with field-effect transistors so it has a low noise level (15 dB per DIN IEC 651), high reliability, and lifelong stability. Low current consumption at low voltage and phantom circuit powering permit feeding the microphone supply voltage via a standard two-conductor shielded audio cable or an internal AA battery.

The K6 offers interchangeable capsules, allowing the selection of different response characteristics from omnidirectional to cardioid to shotgun to adapt the microphone to various types of environments and recording applications.

Because of the new PCM recorders, signal-to-noise ratio (SNR) has reached a level of 90 dB, requiring capacitor microphones to increase their SNR level to match the recorder. The shotgun K6 series microphone, Fig. 1.42, has an equivalent noise level of 16 dB (DIN IEC 651).

FIGURE 1.41   Modular microphone system with an omnidirectional cartridge utilizing a voltage divider circuit. Courtesy Sennheiser Electronic Corporation.

FIGURE 1.42 The same microphone shown in Fig. 1.41 but with a shotgun cartridge. Courtesy Sennheiser Electronic Corporation.

As in most circuitry, the input stage of a voltage divider-type capacitor microphone has the most effect on noise, Fig. 1.43. It is important that the voltage on the transducer does not change. This is normally accomplished by controlling the input current. In the circuit of Fig. 1.43, the voltages V_{in}, V_o, and V_D are within 0.1% of each other. Noise, which might come into the circuit as V_{in} through the operational amplifier, is only $\frac{1}{91}$ of the voltage V_o.

Preattenuation, that is, attenuation between the capacitor and the amplifier, can be achieved by connecting parallel capacitors to the input, by reducing the input stage gain by means of capacitors in the negative feedback circuit, or by reducing the polarizing voltage to one-third its normal value and by using a resistive voltage divider in the audio line. Figure 1.44 is the schematic for the AKG C-460B microphone.

1.3.4.2 Phantom Power for Capacitor Microphones

A common way to supply power for capacitor microphones is with a *phantom circuit*. *Phantom* or *simplex powering* is supplying power to the microphone from the input of the following device such as a preamplifier, mixer, or console.

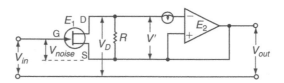

FIGURE 1.43   Simplified schematic of an AKG C-460B microphone input circuit. Courtesy AKG Acoustics.

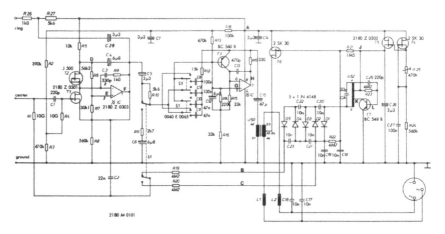

FIGURE 1.44   Schematic of an AKG C-460B microphone. Courtesy AKG Acoustics.

Most capacitor microphone preamplifiers can operate on any voltage between 9 Vdc and 52 Vdc because they incorporate an internal voltage regulator. The preamplifier supplies the proper polarizing voltage for the capacitor capsule plus impedance matches the capsule to the balanced low-impedance output.

Standard low-impedance, balanced microphone input receptacles are easily modified to simplex both operating voltage and audio output signal, offering the following advantages in reduced cost and ease of capacitor microphone operation:

- Special external power supplies and separate multiconductor cables formerly required with capacitor microphones can be eliminated.
- The B+ supply in associated recorders, audio consoles, and commercial sound amplifiers can be used to power the microphone directly.
- Dynamic, ribbon, and capacitor microphones can be used interchangeably on standard, low-impedance, balanced microphone circuits.
- Dynamic, ribbon, and self-powered capacitor microphones may be connected to the modified amplifier input without defeating the microphone operating voltage.
- Any recording, broadcast, and commercial installation can be inexpensively upgraded to capacitor microphone operation using existing, two-conductor microphone cables and electronics.

Phantom circuit use requires only that the microphone operating voltage be applied equally to pins 2 and 3 of the amplifier low-impedance (normally an XLR input) receptacle. Pin 1 remains ground and circuit voltage minus. The polarity of standard microphone cable wiring is not important except for the usual audio polarity requirement (see Section 1.5.3). Two equally effective methods of amplifier powering can be used:

1. Connect an amplifier B+ supply of 9–12 V directly to the ungrounded center tap of the microphone input transformer, as shown in Fig. 1.45. A series-dropping resistor is required for voltages between 12 and 52 V. Figure 1.46 is a typical resistor value chart. A chart can be made for any microphone if the current is known for a particular voltage.

2. A two-resistor, artificial center powering circuit is required when the microphone input transformer is not center-tapped, or when input attenuation networks are used across the input transformer primary. Connect a B+ supply of 9–12 V directly to the artificial center of two 332 Ω, 1% tolerance precision resistors, as shown in Fig. 1.47. Any transformer center tap should not be grounded. For voltages between 12 and 52 V, double the chart resistor value of Fig. 1.46.

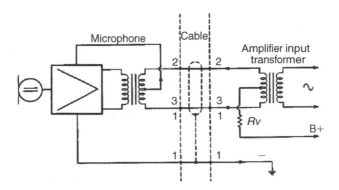

FIGURE 1.45   Direct center-tap connection method of phantom powering capacitor microphones. Courtesy AKG Acoustics.

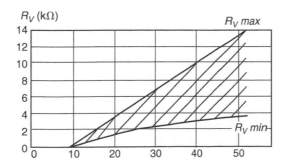

FIGURE 1.46   Dropping resistor value chart for phantom powering AKG C-451E microphones. Courtesy AKG Acoustics.

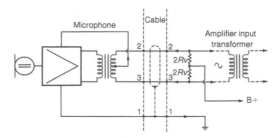

FIGURE 1.47 Artificial center tap connected method of powering capacitor microphones. Courtesy AKG Acoustics.

Any number of capacitor microphones may be powered by either method from a single B+ source according to the current available. Use the largest resistor value shown (R_v *max*) for various voltages in Fig. 1.46 for minimum current consumption.

1.3.4.3 Capacitor Radio-Frequency, Frequency-Modulated Microphones

A frequency-modulated microphone is a capacitor microphone that is connected to a radio-frequency (RF) oscillator. Pressure waves striking the diaphragm cause variations in the capacity of the microphone head that frequency modulates the oscillator. The output of the modulated oscillator is passed to a discriminator and amplified in the usual manner.

Capacitor microphones using an RF oscillator are not entirely new to the recording profession, but since the advent of solid-state devices, considerable improvement has been achieved in design and characteristics. An interesting microphone of this design is the Schoeps Model CMT26U manufactured in West Germany by Schall-Technik, and named after Dr. Carl Schoeps, the designer.

The basic circuitry is shown in Fig. 1.48. By means of a single transistor, two oscillatory circuits are excited and tuned to the exact same frequency of 3.7 MHz. The output voltage from the circuits is rectified by a phase-bridge detector circuit, which operates over a large linear modulation range with very small RF voltages from the oscillator. The amplitude and polarity of the output voltage from the bridge depend on the phase angle between the two high-frequency voltages. The microphone capsule (head) acts as a variable capacitance in one of the oscillator circuits. When a sound wave impinges on the surface of the diaphragm of the microphone head, the vibrations of the diaphragm are detected by the phase curve of the oscillator circuit, and an audio frequency voltage is developed at the output of the bridge circuit. The microphone-head diaphragm is metal to guarantee a large constant capacitance. An automatic frequency control (afc) with a large range of operation is provided

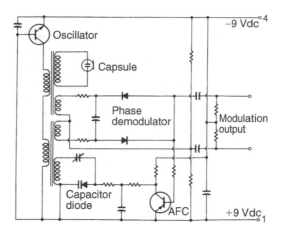

FIGURE 1.48 Basic circuit for the Schoeps radio-frequency capacitor microphone, series CMT.

by means of capacitance diodes to preclude any influence caused by aging or temperature changes on the frequency-determining elements, that might throw the circuitry out of balance.

Internal output resistance is about $200\,\Omega$. The signal, fed directly from the bridge circuit through two capacitors, delivers an output level of $-51\,dB$ to $-49\,dB$ (depending on the polar pattern used) into a $200\,\Omega$ load for a sound pressure level of 10 dynes/cm^2. The SNR and the distortion are independent of the load because of the bridge circuit; therefore, the microphone may be operated into load impedances ranging from 30 to $200\,\Omega$.

1.3.4.3.1 Capacitor Radio-Frequency Microphones

A capacitor microphone of somewhat different design, manufactured by Sennheiser and also employing a crystal-controlled oscillator, is shown in Fig. 1.49. In the conventional capacitor microphone (without an oscillator) the input impedance of the preamplifier is in the order of $100\,M\Omega$ so it is necessary to place the capacitor head and preamplifier in close proximity. In the Sennheiser microphone, the capacitive element (head) used with the RF circuitry is a much lower impedance since the effect of a small change in capacitance at radio frequencies is considerably greater than at audio frequencies. Instead of the capacitor head being subjected to a high dc polarizing potential, the head is subjected to an RF voltage of only a few volts. An external power supply of 12 Vdc is required.

Referring to Fig. 1.49, the output voltage of the 8 MHz oscillator is periodically switched by diodes D_1 and D_2 to capacitor C. The switching phase is shifted 90° from that of the oscillator by means of loose coupling and individually aligning the resonance of the microphone circuit M under a no-sound condition. As a result, the voltage across capacitor C is zero. When a sound impinges on the diaphragm, the switching phase changes proportionally to the sound

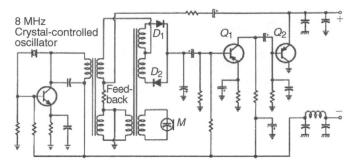

FIGURE 1.49   Basic circuit for the Sennheiser model 105, 405, and 805 capacitor microphones. Courtesy Sennheiser Electronic Corporation.

pressure, and a corresponding audio voltage appears across capacitor C. The output of the switching diodes is directly connected to the transistor amplifier stage, whose gain is limited to 12 dB by the use of negative feedback.

A high Q oscillator circuit is used to eliminate the effects of RF oscillator noise as noise in an oscillatory circuit is inversely proportional to the Q of the circuit. Because of the high Q of the crystal and its stability, compensating circuits are not required, resulting in low internal noise.

The output stage is actually an impedance-matching transformer adjusted for 100 Ω, for a load impedance of 2000 Ω or greater. RF chokes are connected in the output circuit to prevent RF interference and also to prevent external RF fields from being induced into the microphone circuitry.

1.3.4.3.2 Symmetrical Push-Pull Transducer Microphone

Investigations on the linearity of condenser microphones customarily used in the recording studios were carried out by Sennheiser using the *difference frequency method* incorporating a twin tone signal, Fig. 1.50. This is a very reliable test method as the harmonic distortions of both loudspeakers that generate the test sounds separately do not disturb the test result. Thus, difference frequency signals arising at the microphone output are arising from nonlinearities of the microphone itself.

Figure 1.51 shows the distortion characteristics of eight unidirectional studio condenser microphones which were stimulated by two sounds of 104 dB SPL (3 Pa). The frequency difference was fixed to 70 Hz while the twin tone signal was swept through the upper audio range. The curves show that unwanted difference frequency signals of considerable levels were generated by all examined microphones. Although the curves are shaped rather individually, there is a general tendency for increased distortion levels (up to 1% and more) at high frequencies.

The measurement results can be extended to higher signal levels simply by linear extrapolation. This means, for instance, that 10 times higher sound

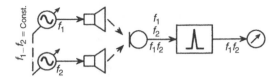

FIGURE 1.50   Difference frequency test. Courtesy Sennheiser Electronic Corporation.

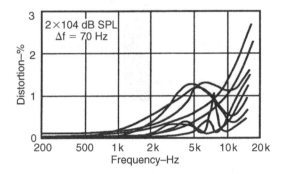

FIGURE 1.51   Frequency distortion of eight unidirectional microphones. Courtesy Sennheiser Electronic Corporation.

pressures will yield 10 times higher distortions, as long as clipping of the microphone circuit is prevented. Thus, two sounds of 124 dB SPL will cause more than 10% distortion in the microphones. Sound pressure levels of this order are beyond the threshold of pain of human hearing but may arise at close-up micing. Despite the fact that the audibility of distortions depends significantly on the tonal structure of the sound signals, distortion figures of this order will considerably affect the fidelity of the sound pickup.

The Cause of Nonlinearity.   Figure 1.52 shows a simplified sketch of a capacitive transducer. The diaphragm and backplate form a capacitor, the capacity of which depends on the width of the air gap. From the acoustical point of view the air gap acts as a complex impedance. This impedance is not constant but depends on the actual position of the diaphragm. Its value is increased if the diaphragm is moved toward the backplate and it is decreased at the opposite movement, so the air gap impedance is varied by the motion of the diaphragm. This implies a parasitic rectifying effect superimposed to the flow of volume velocity through the transducer, resulting in nonlinearity-created distortion.

Solving the Linearity Problem.   A push-pull design of the transducer helps to improve the linearity of condenser microphones, Fig. 1.53. An additional plate equal to the backplate is positioned symmetrically in front of the diaphragm,

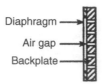

FIGURE 1.52   Conventional capacitor microphone transducer.

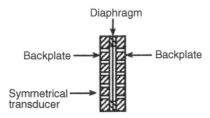

FIGURE 1.53   Symmetrical capacitor microphone transducer.

so two air gaps are formed with equal acoustical impedances as long as the diaphragm is in its rest position. If the diaphragm is deflected by the sound signal, then both air gap impedances are deviated opposite to each other. The impedance of one side increases while the other impedance decreases. The variation effects compensate each other regardless of the direction of the diaphragm motion, and the total air gap impedance is kept constant, reducing the distortion of a capacitive transducer.

Figure 1.54 shows the distortion characteristics of the Sennheiser MKH series push-pull element transformer less RF condenser microphones. The improvement on linearity due to the push-pull design can be seen by comparing Fig. 1.51 to Fig. 1.54.

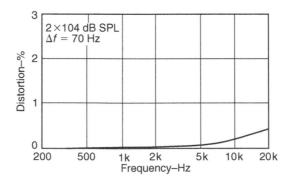

FIGURE 1.54   Distortion characteristics of the symmetrical capacitor microphone transducer.

1.3.4.3.3 Noise Sources

The inherent noise of condenser microphones is caused partly by the random incidence of the air particles at the diaphragm due to their thermal movement. The laws of statistics imply that sound pressure signals at the diaphragm can be evaluated by a precision that improves linearly with the diameter of the diaphragm. Thus, larger diaphragms yield better noise performance than smaller ones.

Another contribution of noise is the frictional effects in the resistive damping elements of the transducer. The noise generation from acoustical resistors is based on the same principles as the noise caused by electrical resistors, so high acoustical damping implies more noise than low damping.

Noise is also added by the electrical circuit of the microphone. This noise contribution depends on the sensitivity of the transducer. High transducer sensitivity reduces the influence of the circuit noise. The inherent noise of the circuit itself depends on the operation principle and on the technical quality of the electrical devices.

Noise Reduction. Large-diameter diaphragms improve noise performance. Unfortunately, a large diameter increases the directivity at high frequencies. A 1 inch (25 mm) transducer diameter is usually a good choice.

A further method to improve the noise characteristics is the reduction of the resistive damping of the transducer. In most directional condenser microphones, a high amount of resistive damping is used in order to realize a flat frequency response of the transducer itself. With this design the electrical circuit of the microphone is rather simple. However, it creates reduced sensitivity and increased noise.

Keeping the resistive damping of the transducer moderate will be a more appropriate method to improve noise performance; however, it leads to the transducer frequency response that is not flat so equalization has to be applied by electrical means to produce a flat frequency response of the complete microphone. This design technique requires a more sophisticated electrical circuit but produces good noise performance.

The electrical output of a transducer acts as a pure capacitance. Its impedance decreases as the frequency increases so the transducer impedance is low in an RF circuit but high in an AF circuit. Moreover, in an RF circuit the electrical impedance of the transducer does not depend on the actual audio frequency but is rather constant due to the fixed frequency of the RF oscillator. Contrary to this, in an AF design, the transducer impedance depends on the actual audio frequency, yielding very high values especially at low frequencies. Resistors of extremely high values are needed at the circuit input to prevent loading of the transducer output. These resistors are responsible for additional noise contribution.

The RF circuit features a very low output impedance which is comparable to that of dynamic-type microphones. The output signal can be applied directly to bipolar transistors, yielding low noise performance by impedance matching.

The Sennheiser MKH 20, Fig. 1.55, is a pressure microphone with omnidirectional characteristics. The MKH 30 is a pure pressure-gradient microphone with a highly symmetrical bidirectional pattern due to the symmetry of the push-pull transducer. The MKH 40, Fig. 1.56, operates as a combined pressure and pressure-gradient microphone yielding a unidirectional cardioid pattern.

- The microphones are phantom powered by 48 Vdc and 2 mA. The outputs are transformer less floated, Fig. 1.57.
- The SPL_{max} is 134 dB at nominal sensitivity and 142 dB at reduced sensitivity.
- The equivalent SPL of the microphones range from 10–12 dBA corresponding to CCIR-weighted figures of 20–22 dB.
- The directional microphones incorporate a switchable bass roll-off to cancel the proximity effect at close-up micing. The compensation is adjusted to about 5 cm (2 in) distance.

A special feature of the omnidirectional microphone is a switchable diffuse field correction that corrects for both direct and diffuse sound field conditions.

FIGURE 1.55   Omnidirectional pressure capacitor microphone. Note the lack of rear entry holes in the case. Courtesy Sennheiser Electronic Corporation.

FIGURE 1.56   Unidirectional pressure/pressure-gradient capacitor microphone. Courtesy Sennheiser Electronic Corporation.

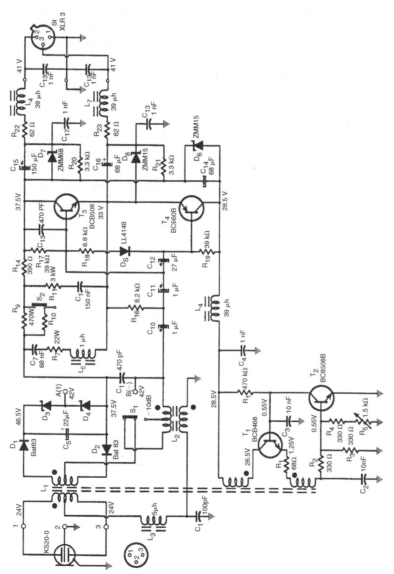

FIGURE 1.57 Schematic of a Sennheiser MKH 20 P 48 U 3 capacitor microphone. Courtesy Sennheiser Electronic Corporation.

The normal switch position is recommended for a neutral pickup when closeup micing and the diffuse field position is used if larger recording distances are used where reverberations become significant.

The distinction between both recording situations arises because omni-directional microphones tend to attenuate lateral and reverse impinging sound signals at high frequencies. Diffuse sound signals with random incidence cause a lack of treble response, which can be compensated by treble emphasis at the microphone. Unfortunately, frontally impinging sounds are emphasized also, but this effect is negligible if the reverberant sound is dominant.

1.3.5 Electret Microphones

An *electret microphone* is a capacitor microphone in which the head capacitor is permanently charged, eliminating the need for a high-voltage bias supply.

From a design viewpoint a microphone intended to be used for critical recording, broadcast, or sound reinforcement represents a challenge involving minimal performance compromise. Early electrets offered the microphone designer a means of reducing the complexity of a condenser microphone by eliminating the high-voltage bias supply, but serious environmental stability problems negated this advantage. Well-designed electret microphones can be stored at 50°C (122°F) and 95% relative humidity for years with a sensitivity loss of only 1 dB. Under normal conditions of temperature and humidity, electret transducers will demonstrate a much lower charge reduction versus time than under the severe conditions indicated. Even if a proper electret material is used, there are many steps in the fabricating, cleaning, and charging processes that greatly influence charge stability.

FIGURE 1.58   Shure SM81 electret capacitor microphone. Courtesy Shure Incorporated.

The Shure SM81 cardioid condenser microphone[3] Fig. 1.58, uses an electret material as a means of establishing a bias voltage on the transducer. The back-plate carries the electret material based upon the physical properties of halo-carbon materials such as Teflon™ and Aclar, which are excellent electrets, and materials such as polypropylene and polyester terephthalate (Mylar™), which are more suitable for diaphragms.

The operation of the Shure SM81 microphone is explained in Section 1.2.3.1.3.

1.4 MICROPHONE SENSITIVITY[7]

Microphone sensitivity is the measure of the electrical output of a microphone with respect to the acoustic sound pressure level input.

Sensitivity is measured in one of three methods:

Open-circuit voltage	$0\,dB = 1\,V/\mu bar$
Maximum power output	$0\,dB = 1\,mW/10\,\mu bar$
	$= 1\,mW/Pa$
Electronic Industries	$0\,dB = EIA$ standard
Association (EIA) sensitivity	SE-105

The common sound pressure levels used for measuring microphone sensitivity are:

94 dB SPL 10 dyn/cm^2 SPL	10 μbar or 1 Pa
74 dB SPL 1 dyn/cm^2 SPL	1 μbar or 0.1 Pa
0 dB SPL 0.0002 dyn/cm^2 SPL	0.0002 Pa or 20 μPa—threshold of hearing

94 dB SPL is recommended since 74 dB SPL is too close to typical noise levels.

1.4.1 Open-Circuit Voltage Sensitivity

There are several good reasons for measuring the open-circuit voltage:

- If the open-circuit voltage and the microphone impedance are known, the microphone performance can be calculated for any condition of loading.
- It corresponds to an effective condition of use. A microphone should be connected to a high impedance to yield maximum SNR. A 150–250 Ω microphone should be connected to 2 kΩ or greater.
- When the microphone is connected to a high impedance compared to its own, variations in microphone impedance do not cause variations in response.

The open-circuit voltage sensitivity (S_v) can be calculated by exposing the microphone to a known SPL, measuring the voltage output, and using the following equation:

$$S_v = 20\log E_o - dB_{SPL} + 94 \tag{1.4}$$

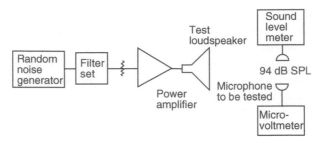

FIGURE 1.59   Method of determining open-circuit voltage sensitivity of a microphone. (From Reference 7.)

where,

S_v is the open-circuit voltage sensitivity in decibels re 1 V for a 10 dyn/cm^2 SPL (94 dB SPL) acoustic input to the microphone,

E_o is the output of the microphone in volts,

dB_{SPL} is the level of the actual acoustic input.

The microphone surement system can be set up as shown in Fig. 1.59. The setup requires a random-noise generator, a microvoltmeter, a high-pass and a low-pass filter set, a power amplifier, a test-loudspeaker, and a sound level meter (SLM). The SLM is placed a specific measuring distance (about 5–6 ft or 1.5–2 m) in front of the loudspeaker. The system is adjusted until the SLM reads 94 dB SPL (a band of pink noise from 250 to 5000 Hz is excellent for this purpose). The microphone to be tested is now substituted for the SLM.

It is often necessary to know the voltage output of the microphone for various SPLs to determine whether the microphone will overload the preamplifier circuit or the SNR will be inadequate. To determine this, use

$$E_o = 10\left(\frac{S_v + dB_{SPL} - 94}{20}\right) \tag{1.5}$$

where,

E_o is the voltage output of microphone,

S_v is the open-circuit voltage sensitivity,

dB_{SPL} is the sound pressure level at the microphone.

1.4.2 Maximum Power Output Sensitivity[7]

The *maximum power output sensitivity* form of specification gives the maximum power output in decibels available from the microphone for a given sound pressure and power reference. Such a specification can be calculated from the internal impedance and the open-circuit voltage of the microphone. This specification also indicates the ability of a microphone to convert sound energy into electrical power. The equation is

$$S_p = 10 \log \frac{V_o^2}{R_o} + 44 \, \text{dB'} \tag{1.6}$$

where,

S_p is the power level microphone sensitivity in decibels,

V_o is the open-circuit voltage produced by a 1 µbar (0.1 Pa) sound pressure,

R_o is the internal impedance of the microphone.

The form of this specification is similar to the voltage specification except that a power as opposed to a voltage reference is given with the sound pressure reference. A 1 mW power reference and a 10 µbar (1 Pa) pressure reference are commonly used (as for the previous case). This form of microphone specification is quite meaningful because it takes into account both the voltage output and the internal impedance of the microphone.

S_p can also be calculated easily from the open circuit voltage sensitivity

$$S_p = S_v - 10 \log Z + 44 \, \text{dB} \tag{1.7}$$

where,

S_p is the decibel rating for an acoustical input of 94 dB_{SPL} (10 dyn/cm^2) or 1 Pa,

Z is the measured impedance of the microphone (the specifications of most manufacturers use the rated impedance).

The output level can also be determined directly from the open-circuit voltage

$$S_p = 10 \log \frac{E_o^2}{0.001Z} - 6 \, \text{dB} \tag{1.8}$$

where,

E_o is the open-circuit voltage,

Z is the microphone impedance.

Because the quantity $10 \log(E^2/0.001Z)$ treats the open-circuit voltage as if it appears across a load, it is necessary to subtract 6 dB. (The reading is 6 dB higher than it would have been had a load been present.)

1.4.3 Electronic Industries Association (EIA) Output Sensitivity

The Electronic Industries Association (EIA) Standard SE-105, August 1949, defines the system rating (G_M) as the ratio of the maximum electrical output from the microphone to the square of the undisturbed sound field pressure in a plane progressive wave at the microphone in decibels relative to 1 mW/0.0002 dyn/cm^2. Expressed mathematically,

$$G_M = 20 \log \frac{E_o}{P} - 10 \log Z_o - 50 \, \text{dB} \tag{1.9}$$

where,

E_o is the open-circuit voltage of the microphone,

P is the undisturbed sound field pressure in dyn/cm^2,

Z_o is the microphone-rated output impedance in ohms.

For all practical purposes, the output level of the microphone can be obtained by adding the sound pressure level relative to 0.0002 dyn/cm^2 to G_M.

Because G_M, S_V, and S_P are compatible, G_M can also be calculated

$$G_M = S_v - 10\log R_{MR} - 50 \text{ dB} \qquad (1.10)$$

where,

G_M is the EIA rating,

R_{MR} is the EIA center value of the nominal impedance range shown below.

Ranges (ohms)		Values Used (ohms)
20–80	=	38
80–300	=	150
300–1250	=	600
1250–4500	=	2400
4500–20,000	=	9600
20,000–70,000	=	40,000

The EIA rating can also be determined from the chart in Fig. 1.60.

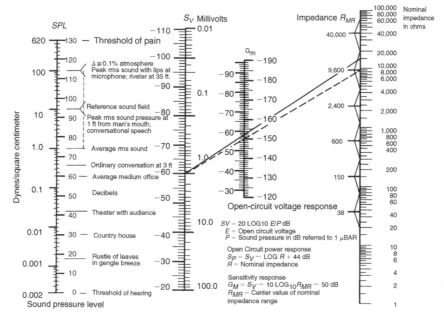

FIGURE 1.60   Microphone sensitivity conversion chart.

1.4.4 Various Microphone Sensitivities

Microphones are subjected to sound pressure levels anywhere from 40 dB SPL when distant micing to 150 dB SPL when extremely close micing (i.e., ¼ inch from the rock singer's mouth or inside a drum or horn).

Various types of microphones have different sensitivities, which is important to know if different types of microphones are intermixed since gain settings, SNR, and preamplifier overload will vary. Table 1.1 gives the sensitivities of a variety of different types of microphones.

TABLE 1.1 Sensitivities of Various Types of Microphones

Type of Microphone	Sp	Sv
Carbon-button	−60 to −50 dB	
Crystal		−50 to −40 dB
Ceramic		−50 to −40 dB
Dynamic (moving coil)	−60 to −52 dB	−85 to −70 dB
Capacitor	−60 to −37 dB	−85 to −45 dB
Ribbon-velocity	−60 to −50 dB	−85 to −70 dB
Transistor	−60 to −40 dB	
Sound power	−32 to −20 dB	
Line level	−40 to 0 dB	−20 to 0 dB
Wireless	−60 to 0 dB	−85 to 0 dB

1.4.5 Microphone Thermal Noise

Since a microphone has an impedance, it generates thermal noise. Even without an acoustic signal, the microphone will still produce a minute output voltage. The thermal noise voltage, E_n, produced by the electrical resistance of a sound source is dependent on the frequency bandwidth under consideration, the magnitude of the resistance, and the temperature existing at the time of the measurement. This voltage is

$$E_n = 4ktR\,(bw) \tag{1.11}$$

where,
k is the Boltzmann's constant, 1.38×10^{-23} J/K,
t is the absolute temperature, $273°$ + room temperature, both in °C,
R is the resistance in ohms,
bw is the bandwidth in hertz.

To change this to dBv use

$$EIN_{dBv} = 20 \log \frac{E_n}{0.775} \qquad (1.12)$$

The thermal noise relative to 1 V is −198 dB for a 1 Hz bandwidth and 1 Ω impedance. Therefore,

$$\frac{TN}{1V} = -198\,dB + 10 \log (bw) + 10 \log Z \qquad (1.13)$$

where,
TN is the thermal noise relative to 1 V,
bw is the bandwidth in hertz,
Z is the microphone impedance in ohms.

Thermal noise relative to 1 V can be converted to equivalent input noise (EIN) by

$$EIN_{dBm} = -198\,dB + 10 \log (bw) + 10 \log Z - 6 - 20 \log 0.775\,V \qquad (1.14)$$

Since the EIN is in dBm and dBm is referenced to 600 Ω, the impedance Z is 600 Ω.

1.5 MICROPHONE PRACTICES

1.5.1 Placement

Microphones are placed in various relationships to the sound source to obtain various sounds. Whatever position gives the desired effect that is wanted is the correct position. There are no exact rules that must be followed; however, certain recommendations should be followed to assure a good sound.

1.5.1.1 Microphone-to-Source Distance

Microphones are normally used in the direct field. Under this condition, inverse square law attenuation prevails, meaning that each time the distance is doubled, the microphone output is reduced 6 dB. For instance, moving from a microphone-to-source distance of 2.5 to 5 cm (1 to 2 in) has the same effect as moving from 15 to 30 cm (6 to 12 in), 1 to 2 ft (30 to 60 cm), or 5 to 10 ft (1.5 to 3 m).

Distance has many effects on the system. In a reinforcement system, doubling the distance reduces gain before feedback 6 dB; in all systems, it reduces the effect of microphone-to-source variations.

Using the inverse-square-law equation for attenuation,

$$attenuation_{dB} = 20 \log \frac{D_1}{D_2} \qquad (1.15)$$

it can be seen, at a microphone-to-source distance of 2.5 cm (1 in), moving the microphone only 1.25 cm (½ in) closer will increase the signal 6 dB and 1.25 cm (½ in) farther away will decrease the signal 3.5 dB for a total signal variation of 9.5 dB for only 2.5 cm (1 in) of total movement! At a source-to-microphone distance of 30 cm (12 in), a movement of 2.5 cm (1 in) will cause a signal variation of only 0.72 dB. Both conditions can be used advantageously; for instance, close micing is useful in feedback-prone areas, high noise level areas (rock groups), or where the talent wants to use the source-to-microphone variations to create an effect.

The farther distances are most useful where lecterns and table microphones are used or where the talker wants movement without level change.

The microphone-to-source distance also has an effect on the sound of a microphone, particularly one with a cardioid pattern. As the distance decreases, the proximity effect increases creating a bassy sound (see Section 1.2.3.1). Closing in on the microphone also increases breath noise and pop noise.

1.5.1.2 Distance from Large Surfaces

When a microphone is placed next to a large surface such as the floor, 6 dB of gain can be realized, which can be a help when far micing.

As the microphone is moved away from the large surface but still in proximity of it, cancellation of some specific frequencies will occur, creating a notch of up to 30 dB, Fig. 1.65. The notch is created by the cancellation of a frequency that, after reflecting off the surface, reaches the microphone diaphragm 180° out of polarity with the direct sound.

The frequency of cancellation, f_c, can be calculated from the equation

$$f_c = \frac{0.5c}{D_{r1} + D_{r2} - D_d} \tag{1.16}$$

where,
c is the speed of sound, 1130 feet per second or 344 meters per second,
0.5 is the out-of-polarity frequency ratio,
D_{r1} is the reflected path from the source to the surface in feet or meters,
D_{r2} is the reflected path from the surface to the microphone in feet or meters,
D_d is the direct path from the source to the microphone in feet or meters.

If the microphone is 10 ft from the source and both are 5 ft above the floor, the canceled frequency is

$$\begin{aligned} f_c &= \frac{1130 \times 0.5}{7.07 + 7.07 - 10} \\ &= 136.47 \, \text{Hz} \end{aligned} \tag{1.17}$$

If the microphone is moved to 2 ft above the floor, the canceled frequency is 319.20 Hz. If the microphone is 6 inches from the floor, the canceled frequency is 1266.6 Hz. If the microphone is 1 inch from the floor, the canceled frequency is 7239.7 Hz.

1.5.1.3 Behind Objects

Sound, like light, does not go through solid or acoustically opaque objects. It does, however, go through objects of various density. The transmission loss or ability of sound to go through this type of material is frequency dependent; therefore, if an object of this type is placed between the sound source and the microphone, the pickup will be attenuated according to the transmission characteristics of the object.

Low-frequency sound bends around objects smaller than their wavelength, which affects the frequency response of the signal. The normal effect of placing the microphone behind an object is an overall reduction of level, a low-frequency boost, and a high-frequency roll-off.

1.5.1.4 Above the Source

When the microphone is placed above or to the side of a directional sound source (i.e., horn or trumpet), the high-end frequency response will roll off because high frequencies are more directional than low frequencies, so less high-frequency SPL will reach the microphone than low-frequency SPL.

1.5.1.5 Direct versus Reverberant Field

Micing in the reverberant field picks up the characteristic of the room because the microphone is picking up as much or more of the room, as it is the direct sound from the source. When micing in the reverberant field, only two microphones are required for stereo since isolation of the individual sound sources is impossible. When in the reverberant field, a directional microphone will lose much of its directivity. Therefore, it is often advantageous to use an omnidirectional microphone that has smoother frequency response. To mic sources individually, you must be in the direct field and usually very close to the source to eliminate cross-feed.

1.5.2 Grounding

The grounding of microphones and their interconnecting cables is of extreme importance since any hum or noise picked up by the cables will be amplified along with the audio signal. Professional systems generally use the method shown in Fig. 1.61. Here the signal is passed through a two-conductor shielded cable to the balanced input of a preamplifier. The cable shield is connected to pin number 1, and the audio signal is carried by the two conductors and pins 2 and 3 of the XLR-type connector. The actual physical ground is connected at the preamplifier chassis only and carried to the microphone case. In no instance is a second ground ever connected to the far end of the cable, because this will cause the flow of ground currents between two points of grounding.

In systems designed for semiprofessional and home use, the method in Fig. 1.62 is often used. Note that one side of the audio signal is carried over the

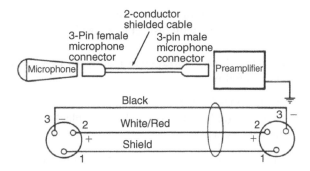

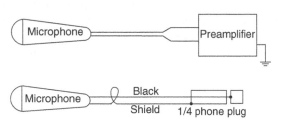

FIGURE 1.61   Typical low-impedance microphone to preamplifier wiring.

FIGURE 1.62   Typical semiprofessional, hi-fi microphone to preamplifier wiring.

cable shield to a pin-type connector. The bodies of both the male and female connector are grounded: the female to the amplifier case and the male to the cable shield. The microphone end is connected in a similar manner; here again the physical ground is connected only at the preamplifier chassis. Hum picked up on the shield and not on the center conductor is added to the signal and amplified through the system.

1.5.3 Polarity

Microphone polarity, or *phase* as it is often called, is important especially when multiple microphones are used. When they are in polarity they add to each other rather than have canceling effects. If multiple microphones are used and one is out of polarity, it will cause comb filters, reducing quality and stereo enhancement. The EIA standard RS-221.A, October 1979, states "Polarity of a microphone or a microphone transducer element refers to in-phase or out-of-phase condition of voltage developed at its terminals with respect to the sound pressure of a sound wave causing the voltage."

Note: *Exact in-phase relationship* can be taken to mean that the voltage is coincident with the phase of the sound pressure wave causing the voltage. In practical microphones, this perfect relationship may not always be obtainable.

The positive or in-phase terminal is that terminal that has a positive potential and a phase angle less than 90° with respect to a positive sound pressure at the front of the diaphragm.

When connected to a *three-pin XLR connector* as per EIA standard RS-297, the polarity shall be as follows:

- Out-of-phase—terminal 3 (black).
- In-phase—terminal 2 (red or any color other than black).
- Ground—terminal 1 (shield).

Figure 1.63 shows the proper polarity for three-pin and five-pin XLR connectors and for three-pin and five-pin DIN connectors.

A simple method of determining microphone polarity is as follows:

If two microphones have the same frequency response and sensitivity and are placed next to each other and connected to the same mixer, the output will double if both are used. However, if they are out of polarity with each other, the total output will be down 40–50 dB from the output of only one microphone.

The microphones to be tested for proper polarity are placed alongside each other and connected to their respective mixer inputs. With a single acoustic source into the microphones (pink noise is a good source), one mixer volume

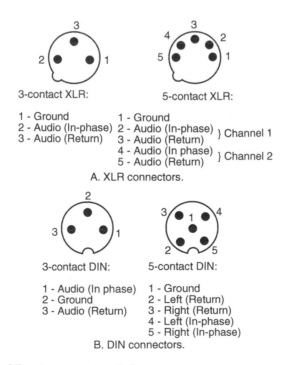

FIGURE 1.63   Microphone connector polarity.

control is adjusted for a normal output level as indicated on a VU meter. Note the volume control setting and turn it off. Make the same adjustment for the second microphone, and note the setting of this volume control. Now open both controls to these settings. If the microphones are out of polarity, the quality of reproduction will be distorted, and there will be a distinct drop in level. Reversing the electrical connections to one microphone will bring them into polarity, making the quality about the same as one microphone operating and the output level higher.

If the microphones are of the bidirectional type, one may be turned 180° to bring it into polarity and later corrected electrically. If the microphones are of the directional type, only the output or cable connections can be reversed. After polarizing a bidirectional microphone, the rear should be marked with a white stripe for future reference.

1.5.4 Balanced or Unbalanced

Microphones can be connected either *balanced* or *unbalanced*. All professional installations use a balanced system for the following reasons:

- Reduced pickup of hum.
- Reduced pickup of electrical noise and transients.
- Reduced pickup of electrical signals from adjacent wires.

These reductions are realized because the two signal conductors shown in Fig. 1.64 pickup the same stray signal with equal intensity and polarity, so the noise is impressed evenly on each end of the transformer primary, eliminating a potential across the transformer and canceling any input noise. Because the balanced wires are in a shielded cable, the signal to each conductor is also greatly reduced.

When installing microphones into an unbalanced system, any noise that gets to the inner unbalanced conductor is not canceled by the noise in the shield, so the noise is transmitted into the preamplifier. In fact, noise impressed on the microphone end of the shield adds to the signal because of the resistance of the shield between the noise and the amplifier.

Balanced low-impedance microphone lines can be as long as 500 ft (150 m) but unbalanced microphone lines should never exceed 15 ft (4.5 m).

1.5.5 Impedance

Most professional microphones are low impedance, $200\,\Omega$, and are designed to work into a load of $2000\,\Omega$. High-impedance microphones are $50,000\,\Omega$ and are designed to work into an impedance of $1–10\,M\Omega$. The low-impedance microphone has the following advantages:

- Less susceptible to noise. A noise source of relatively high impedance cannot "drive" into a source of relatively low impedance (i.e., the microphone cable).

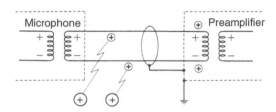

FIGURE 1.64   Noise cancellation on balanced, shielded microphone cables.

- Capable of being connected to long microphone lines without noise pickup and high-frequency loss.

All microphone cable has inductance and capacitance. The capacitance is about 40 pF (40×10^{-12})/ft (30 cm). If a cable is 100 ft long (30 m), the capacitance would be (40×10^{-12}) $\times$ 100 ft or 4×10^{-9} F or 0.004 μF. This is equivalent to a 3978.9 Ω impedance at 10,000 Hz and is found with the equation

$$X_c = \frac{1}{2\pi fC} \tag{1.18}$$

This has little effect on a microphone with an impedance of 200 Ω as it does not reduce the impedance appreciably as determined by

$$Z_r = \frac{X_c Z_m}{X_c + Z_m} \tag{1.19}$$

For a microphone impedance of 200 Ω, the total impedance $Z_T = 190\,\Omega$ or less than 0.5 dB.

If this same cable were used with a high-impedance microphone of 50,000 Ω, 10,000 Hz would be down more than 20 dB.

Making the load impedance equal to the microphone impedance will reduce the microphone sensitivity 6 dB, which reduces the overall SNR by 6 dB. For the best SNR, the input impedance of low-impedance microphone preamplifiers is always 2000 Ω or greater.

If the load impedance is reduced to less than the microphone impedance, or the load impedance is not resistive, the microphone frequency response and output voltage will be affected.

Changing the load of a high-impedance or ceramic microphone from 10 MΩ to 100 kΩ reduces the output at 100 Hz by 27 dB.

1.6 MISCELLANEOUS MICROPHONES

1.6.1 Pressure Zone Microphones (PZM)

The *pressure zone microphone*, referred to as a *PZ Microphone* or *PZM*, is a miniature condenser microphone mounted face-down next to a sound-reflecting plate or boundary. The microphone diaphragm is placed in the pressure zone just above the boundary where direct and reflected sounds combine effectively in-phase over the audible range.

In many recording and reinforcement applications, the sound engineer is forced to place microphones near hard reflective surfaces such as when recording an instrument surrounded by reflective baffles, reinforcing drama or opera with the microphones near the stage floor, or recording a piano with the microphone close to the open lid.

In these situations, sound travels from the source to the microphone via two paths: directly from the source to the microphone, and reflected off the surface to the microphone. The delayed sound reflections combine with the direct sound at the microphone, resulting in phase cancellations of various frequencies, Fig. 1.65. This creates a series of peaks and dips in the net frequency response called the *comb-filter effect*, affecting the recorded tone quality and giving an unnatural sound.

The PZM was developed to avoid the tonal coloration caused by microphone placement near a surface. The microphone diaphragm is arranged parallel with and very close to the reflecting surface and facing it, so that the direct

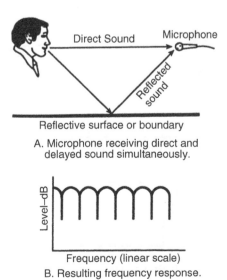

A. Microphone receiving direct and
delayed sound simultaneously.

B. Resulting frequency response.

FIGURE 1.65   Effects of cancellation caused by near reflections (comb filters).

and reflected waves combine at the diaphragm in-phase over the audible range, Fig. 1.66.

This arrangement can provide several benefits:

- Wide, smooth frequency response (natural reproduction) because of the lack of phase interference between direct and reflected sound.
- A 6 dB increase in sensitivity because of the coherent addition of direct and reflected sound.
- High SNR created by the PZM's high sensitivity and low internal noise.
- A 3 dB reduction in pickup of the reverberant field compared to a conventional omnidirectional microphone.
- Lack of off-axis coloration as a result of the sound entry's small size and radial symmetry.
- Good-sounding pickup of off-mic instruments due to the lack of off-axis coloration.
- Identical frequency response for random-incidence sound (ambience) and direct sound due to the lack of off-axis coloration.
- Consistent tone quality regardless of sound-source movement or microphone-to-source distance.
- Excellent reach (clear pickup of quiet distant sounds).
- Hemispherical polar pattern, equal sensitivity to sounds coming from any direction above the surface plane.
- Inconspicuous low-profile mounting.

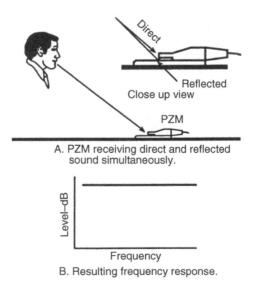

FIGURE 1.66   Effects of receiving direct and reflected sound simultaneously.

FIGURE 1.67   PZM 30D pressure zone microphone. Courtesy Crown International, Inc.

The Crown PZM-30 series microphones, Fig. 1.67, one of the original PZMs, are designed for professional use and built to take the normal abuse associated with professional applications. Miniaturized electronics built into the microphone cantilever allow the PZM-30 series to be powered directly by simplex phantom powering.

1.6.1.1 Phase Coherent Cardioid (PCC)

The *phase coherent cardioid microphone* (PCC) is a surface-mounted supercardioid microphone with many of the same benefits as the PZM. Unlike the PZM, however, the PCC uses a subminiature supercardioid microphone capsule.

Technically, the PCC is not a pressure zone microphone. The diaphragm of a PZM is parallel to the boundary; the diaphragm of the PCC is perpendicular to the boundary. Unlike a PZM, the PCC aims along the plane on which it is mounted. In other words, the main pickup axis is parallel with the plane.

The Crown PCC-200 microphone, a Phase Coherent Cardioid surface-mounted boundary microphone, Fig. 1.68, is intended for use on stage floors, lecterns, and conference tables wherever gain-before-feedback and articulation are important. Figure 1.69 shows the horizontal polar response for this microphone.

The PCC-160 can be directly phantom powered. A bass-tilt switch is provided for tailoring low end response.

1.6.1.2 Directivity

The PZM picks up sounds arriving from any direction above the surface it is mounted on (hemispherical). It is often necessary to discriminate against sounds arriving from certain directions. To make the microphone directional or hemicardioid (reject sounds from the rear) the capsule can be mounted with the cantilever in a corner boundary made of ¼ inch (6 mm) thick Plexiglas. The larger the boundary, the better it discriminates against low-frequency sounds from the rear.

FIGURE 1.68 Crown PCC-200 Phase Coherent Cardioid microphone. Courtesy Crown International, Inc.

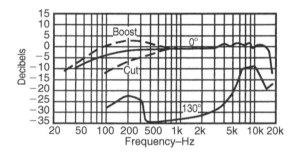

FIGURE 1.69 The horizontal plane polar response of the PCC-160 Phase Coherent Cardioid microphone with the source 30° above the infinite boundary. Courtesy Crown International, Inc.

For best results, a corner boundary 12 in × 24 in wide (0.3 m × 0.6 m) is recommended and is nearly invisible to the audience, Fig. 1.70.

A boom-mounted or suspended PZM can be taped to the center of a ¼ inch (6 mm) thick 2 ft × 2 ft (0.6 m × 0.6 m) or 4 ft × 4 ft (1.2 m × 1.2 m) panel. The microphone should be placed 4 inches (10 cm) off-center for a smoother frequency response. Using clear acrylic plastic (Plexiglas) makes the panel nearly invisible from a distance. If the edges of the Plexiglas pickup light, they can be taped or painted black.

1.6.1.2.1 Sensitivity Effects

If a PZM capsule is placed very near a single large boundary (within 0.020 in or 0.50 mm), such as a large plate, floor, or wall, incoming sound reflects off the surface. The reflected sound wave adds to the incoming sound wave in the

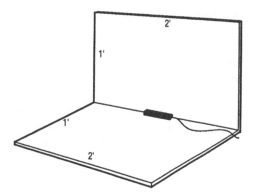

FIGURE 1.70   Corner boundary used to control directivity of pressure zone microphones.

Pressure Zone next to the boundary. This coherent addition of sound waves doubles the sound pressure at the microphone, effectively increasing the microphone sensitivity or output by 6 dB over a standard microphone.

If the PZM capsule is placed at the junction of two boundaries at right angles to each other, such as the floor and a wall, the wall increases sensitivity 6 dB, and the floor increases sensitivity another 6 dB. Adding two boundaries at right angles increases sensitivity 12 dB.

With the PZM element at the junction of three boundaries at right angles, such as in the corner of the floor and two walls, microphone sensitivity will be 18 dB higher than what it was in open space.

Note that the acoustic sensitivity of the microphone rises as boundaries are added, but the electronic noise of the microphone stays constant, so the effective SNR of the microphone improves 6 dB every time a boundary is added at right angles to previous boundaries.

1.6.1.2.2 Direct-to-Reverberant Ratio Effects

Direct sound sensitivity increases 6 dB per boundary added, while reverberant or random-incidence sound increases only 3 dB per boundary added. Consequently, the direct-to-reverberant ratio increases 3 dB ($6\,dB_{dir} - 3\,dB_{rev}$) whenever a boundary is added at right angles to previous boundaries.

1.6.1.2.3 Frequency-Response Effects

The low-frequency response of the PZM or PCC depends on the size of the surface it is mounted on. The larger the surface, the more the low-frequency response is extended. The low-frequency response shelves down to a level 6 dB below the mid-frequency level at the frequency where the wavelength is about 6 times the boundary dimension. For example, the frequency response of a PZM on a 2 ft × 2 ft (0.6 m × 0.6 m) panel shelves down 6 dB below 94 Hz. On a 5 in × 5 in (12 cm × 12 cm) plate, the response shelves down 6 dB below 376 Hz.

For best bass and flattest frequency response, the PZM or PCC must be placed on a large hard boundary such as a floor, wall, table, or baffle at least 4 ft × 4 ft (1.2 m × 1.2 m).

To reduce bass response, the PZM or PCC can be mounted on a small plate well away from other reflecting surfaces. This plate can be made of thin plywood, Masonite, clear plastic, or any other hard, smooth material. When used on a carpeted floor the PZM or PCC should be placed on a hard-surfaced panel at least 1 ft × 1 ft (0.3 m × 0.3 m) for flattest high-frequency response.

To determine the frequency $f_{-6\,dB}$ where the response shelves down 6 dB, use

$$f_{-6\,dB} = \frac{188*}{D} \tag{1.20}$$

*57.3 for SI units
where,
D is the boundary dimension in feet or meters.

For example, if the boundary is 2 ft (0.6 m) square, the 6 dB down point is

$$\begin{aligned} f_{-6\,dB} &= \frac{188}{D} \\ &= \frac{188}{2} \\ &= 94\,Hz \end{aligned}$$

Below 94 Hz, the response is a constant 6 dB below the upper mid-frequency level. Note that there is a response shelf, not a continuous roll-off.

When the PZM is on a rectangular boundary, two shelves appear. The long side of the boundary is D_{max} and the short side D_{min}. The response is down 3 dB at

$$f_{-3\,dB} = \frac{188*}{D_{max}} \tag{1.21}$$

*57.3 for SI units

and is down another 3 dB at

$$f_{-3\,dB} = \frac{188*}{D_{min}} \tag{1.22}$$

*57.3 for SI units

The low-frequency shelf varies with the angle of the sound source around the boundary. At 90° incidence (sound wave motion parallel to the boundary), there is no low-frequency shelf.

The depth of the shelf also varies with the distance of the sound source to the panel. The shelf starts to disappear when the source is closer than a panel dimension away. If the source is very close to the PZM mounted on a panel, there is no low-frequency shelf; the frequency response is flat.

If the PZM is at the junction of two or more boundaries at right angles to each other, the response shelves down 6 dB per boundary at the above frequency. For example, a two-boundary unit made of 2 ft (0.6 m) square panels shelves down 12 dB below 94 Hz.

There are other frequency-response effects in addition to the low-frequency shelf. For sound sources on-axis to the boundary, the response rises about 10 dB above the shelf at the frequency where the wavelength equals the boundary dimension.

For a square panel,

$$F_{peak} = \frac{0.88c}{D} \qquad (1.23)$$

where,
c is the speed of sound (1130 ft/s or 344 m/s) D is the boundary dimension in feet or meters.

For a circular panel

$$F_{peak} = \frac{c}{D} \qquad (1.24)$$

As an example, a 2 ft (0.6 m) square panel has a 10 dB rise above the shelf at

$$\begin{aligned} F_{peak} &= \frac{0.88c}{D} \\ &= \frac{0.88 \times 1130}{2} \\ &= 497\,\text{Hz} \end{aligned}$$

Note that this response peak is only for the direct sound of an on-axis source. The effect is much less if the sound field at the panel is partly reverberant, or if the sound waves strike the panel at an angle. The peak is also reduced if the microphone capsule is placed off-center on the boundary.

Figure 1.71 shows the frequency response at various angles of sound incidence of a PZM mounted on a 2 ft square panel. Note the several phenomena shown in the figure:

- The low-frequency shelf (most visible at 30° and 60°).
- The lack of low-frequency shelving at 90° (grazing incidence).
- The 10 dB rise in response at 497 Hz.
- The edge-interference peaks and dips above 497 Hz (most visible at 0° or normal incidence).
- The lessening of interference at increasing angles.
- The greater rear rejection of high frequencies than low frequencies.

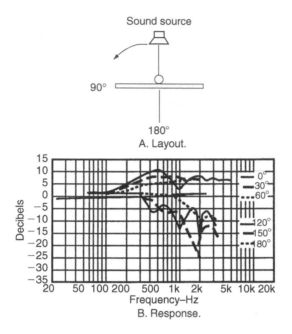

FIGURE 1.71   Frequency response of a pressure zone microphone. Note the 6 dB shelf at 94 Hz.

1.6.1.2.4 Frequency-Response Anomalies Caused by Boundaries

Frequency response is affected by the following:

- When sound waves strike a boundary, pressure doubling occurs at the boundary surface, but does not occur outside the boundary, so there is a pressure difference at the edge of the boundary. This pressure difference creates sound waves.

    These sound waves generated at the edge of the boundary travel to the microphone in the center of the boundary. At low frequencies, these edge waves are opposite in polarity to the incoming sound waves. Consequently, the edge waves cancel the pressure-doubling effect.

- At low frequencies, pressure doubling does not occur, but at mid to high frequencies, pressure doubling does occur. The net effect is a mid- to high-frequency boost, which could be considered a low-frequency loss or shelf.

- Incoming waves having wavelengths about six times the boundary dimensions are canceled by edge effects while waves much smaller than the boundary dimensions are not canceled by edge effects.

- Waves having wavelengths on the order of the boundary dimensions are subject to varying interference versus frequency, i.e., peaks and dips in the frequency response.

- At the frequency where the wavelength equals the boundary dimension, the edge wave is in phase with the incoming wave. Consequently, there is a response rise (about 10 dB above the low-frequency shelf) at that frequency. Above that frequency, there is a series of peaks and dips that decrease in amplitude with frequency.
- The edge-wave interference decreases if the incoming sound waves approach the boundary at an angle.
- Interference also is reduced by placing the microphone capsule off-center. This randomizes the distances from the edges to the microphone capsule, resulting in a smoother response.

1.6.2 Lavalier Microphones

Lavalier microphones are made either to wear on a lavalier around the neck or to clip onto a tie, shirt, or other piece of clothing. The older heavy style lavalier microphone, Fig. 1.72, which actually laid on the chest, had a frequency response that was shaped to reduce the boomingness of the chest cavity, and the loss of high-frequency response caused by being 90° off axis to the signal, Fig. 1.73. These microphones should not be used for anything except as a lavalier microphone.

Lavalier microphones may be dynamic, condenser (capacitor), pressure-zone, electret, or high-impedance ceramic.

The newer loss mass clip-on lavalier microphones, Figs. 1.74 and 1.75, do not require frequency response correction because there is no coupling to the chest cavity and the small diameter of the diaphragm does not create pressure build-up at high frequencies, creating directionality.

Most lavalier microphones are omnidirectional; however, more manufacturers are producing directional lavalier microphones. The Sennheiser MKE 104 clip-on lavalier microphone, Fig. 1.75, has a cardioid pickup pattern, Fig. 1.76. This reduces feedback, background noise, and comb filtering caused by the canceling between the direct sound waves and sound waves that hit the microphone on a reflective path from the floor, lectern, and so forth.

FIGURE 1.72   Shure SM11 dynamic omnidirectional lavalier microphone. Courtesy Shure Incorporated.

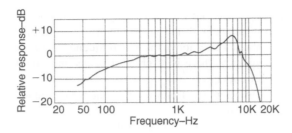

FIGURE 1.73   Typical frequency response of a heavy-style dynamic lavalier microphone.

FIGURE 1.74   Shure SM183 omnidirectional condenser lavalier microphone. Courtesy Shure Incorporated.

FIGURE 1.75   Sennheiser MKE 104 clip-on lavalier directional microphone. Courtesy Sennheiser Electronic Corporation.

One of the smallest microphones is the Countryman B6, Fig. 1.77. The B6 microphone has a diameter of 0.1 inches and has replaceable protective caps. Because of its small size, it can be hidden even when it's in plain sight. By choosing a color cap to match the environment, the microphone can be pushed through a button hole or placed in the hair.

Lavalier microphones are normally used to give the talker freedom of movement. This causes problems associated with motion—for instance, noise being transmitted through the microphone cable. To reduce this noise, soft, flexible microphone cable with good fill to reduce wire movement should be used. The cable, or power supply for electret/condenser microphones, should be clipped to

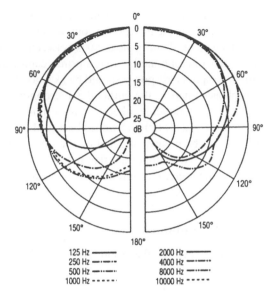

FIGURE 1.76   Polar response of the microphone in Fig. 1.75. Courtesy Sennheiser Electronic Corporation.

FIGURE 1.77   Countryman B6 miniature lavalier microphone. Courtesy of Countryman Associates, Inc.

the user's belt or pants to reduce cable noise to only that created between the clip and the microphone, about 2 ft (0.6 m). Clipping to the waist also has the advantage of acting as a strain relief when the cord is accidentally pulled or stepped on.

A second important characteristic of the microphone cable is size. The cable should be as small as possible to make it unobtrusive and light enough so it will not pull on the microphone and clothing.

Because the microphone is normally 10 inches (25 cm) from the mouth of the talker and out of the signal path, the microphone output is less than a microphone on a stand in front of the talker. Unless the torso is between the microphone and loudspeaker, the lavalier microphone is often a prime candidate for feedback. For this reason, the microphone response should be as smooth as possible.

As in any microphone situation, the farther the microphone is away from the source, the more freedom of movement between microphone and source without adverse effects. If the microphone is worn close to the neck for increased gain, the output level will be greatly affected by the raising and lowering and turning of the talker's head. It is important that the microphone be worn chest high and free from clothing, etc. that might cover the capsule, reducing high-frequency response.

1.6.3 Head-Worn Microphones

Head-worn microphones such as the Shure Model SM10A, Fig. 1.78, and Shure Beta 53, Fig. 1.79, are low-impedance, unidirectional, dynamic microphones, designed for sports and news announcing, for interviewing and inter-communications systems, and for special-event remote broadcasting. The Shure SM10A is a unidirectional microphone while the Beta 53 is an omnidirectional microphone.

Head-worn microphones offer convenient, hands-free operation without user fatigue. As close-talking units, they may be used under noisy conditions

FIGURE 1.78 Shure SM10A dynamic unidirectional head-worn microphone. Courtesy Shure Incorporated.

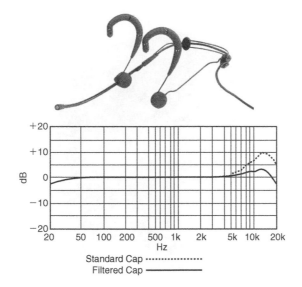

FIGURE 1.79 Shure Beta 53 omnidirectional microphone and its frequency response. Courtesy Shure Incorporated.

without losing or masking voice signals. They are small, lightweight, rugged, and reliable units that normally mount to a cushioned headband. A pivot permits the microphone boom to be moved 20° in any direction and the distance between the microphone and pivot to be changed 9 cm (3½ in).

Another head-worn microphone, the Countryman Isomax E6 Directional EarSet microphone, is extremely small. The microphone clips around the ear rather than around the head. The units come in different colors to blend in with the background. The ultra-miniature condenser element is held close to the mouth by a thin boom and comfortable ear clip. The entire assembly weighs less than one-tenth of an ounce and almost disappears against the skin, so performers can forget it's there and audiences barely see it, Fig. 1.80.

The microphone requires changeable end caps that create a cardioid pickup pattern for ease of placement, or a hypercardioid pattern when more isolation is needed. The C (cardioid) and H (hypercardioid) end caps modify the EarSet's directionality, Fig. 1.81.

The EarSet series should always have a protective cap in place to keep sweat, makeup, and other foreign material out of the microphone.

The hypercardioid cap provides the best isolation from all directions, with a null toward the floor where wedge monitors are often placed. The hypercardioid is slightly more sensitive to air movement and handling noise and should always be used with a windscreen.

FIGURE 1.80   Countryman Isomax E6 Directional EarSet microphone. Courtesy Countryman Associates, Inc.

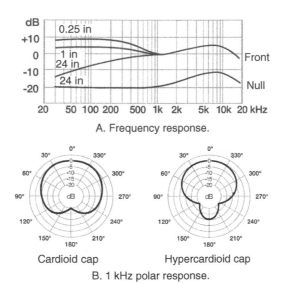

A. Frequency response.

Cardioid cap   Hypercardioid cap

B. 1 kHz polar response.

FIGURE 1.81   Frequency response and polar response of the Countryman E6 EarSet microphone in Fig. 1.80. Courtesy Countryman Associates, Inc.

The cardioid cap is slightly less directional, with a null roughly toward the performer's back. It's useful for trade show presenters or others who have a monitor loudspeaker over their shoulders or behind them.

The microphone can be connected to a sound board or wireless microphone transmitter with the standard 2 mm cable or an extra small 1 mm cable, Fig. 1.80.

1.6.4 Base Station Microphones

Base station power microphones are designed specifically for citizens band transceivers, amateur radio, and two-way radio applications. For clearer transmission and improved reliability, transistorized microphones can be used to replace ceramic or dynamic, high- or low-impedance microphones supplied as original equipment.

The Shure Model 450 Series II, Fig. 1.82, is a high output dynamic microphone designed for paging and dispatching applications. The microphone has an omnidirectional pickup pattern and a frequency response tailored for optimum speech intelligibility, Fig. 1.83. It includes an output impedance selection switch for high, 30,000 Ω, and low, 225 Ω, and a locking press-to-talk switch.

The press-to-talk switch can be converted to a monitor/transmit switch with the Shure RK199S Split-Bar Conversion Kit. When the optional split-bar Transmit/Monitor Switch Conversion Kit is installed, the monitor bar must be depressed before the transmit switch can be depressed, requiring the operator

FIGURE 1.82   Shure 450 Series II base station microphone. Courtesy Shure Incorporated.

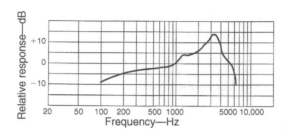

FIGURE 1.83 Frequency response of the Shure 450 Series II microphone shown in Fig. 1.82. Courtesy Shure Incorporated.

to verify that the channel is open before transmitting. The monitor bar can be locked in the on position. The transmit bar is momentary and cannot be locked.

1.6.5 Differential Noise-Canceling Microphones

Differential noise-canceling microphones, Fig. 1.84, are essentially designed for use in automobiles, aircraft, boats, tanks, public-address systems, industrial plants, or any service where the ambient noise level is 80 dB or greater and the microphone is handheld. Discrimination is afforded against all sounds originating more than ¼ inch (6.4 mm) from the front of the microphone. The noise-canceling characteristic is achieved through the use of a balanced port opening, which directs the unwanted sound to the rear of the dynamic unit diaphragm out of phase with the sound arriving at the front of the microphone. The noise canceling is most effective for frequencies above 2000 Hz. Only speech originating within ¼ inch (6.4 mm) of the aperture is fully reproduced. The average discrimination between speech and noise is 20 dB with a frequency response of 200–5000 Hz.

FIGURE 1.84   Shure 577B dynamic noise-canceling microphone. Courtesy Shure Incorporated.

1.6.6 Controlled-Reluctance Microphones

The *controlled-reluctance microphone* operates on the principle that an electrical current is induced in a coil, located in a changing magnetic field. A magnetic armature is attached to a diaphragm suspended inside a coil. The diaphragm, when disturbed by a sound wave, moves the armature and induces a corresponding varying voltage in the coil. High output with fairly good frequency response is typical of this type of microphone.

1.6.7 Handheld Entertainer Microphones

The *handheld entertainer microphone* is most often used by a performer on stage and, therefore, requires a special frequency response that will increase articulation and presence. The microphones are often subjected to rough handling, extreme shock, and vibration. For live performances, the proximity effect can be useful to produce a low bass sound.

Probably the most famous entertainer's microphone is the Shure SM58, Fig. 1.85. The microphone has a highly effective spherical windscreen that also reduces breath pop noise. The cardioid pickup pattern helps reduce feedback. The frequency response, Fig. 1.86, is tailored for vocals with brightened midrange and bass roll-off. Table 1.2 gives the suggested microphone placement for best tone quality.

To overcome rough handling and handling noise, special construction techniques are used to reduce wind, pop noise, and mechanical noise and to ensure that the microphone will withstand sudden collisions with the floor. The Sennheiser MD431, Fig. 1.87, is an example of a high-quality, rugged, and low-mechanical-noise microphone. To eliminate feedback, the MD431 incorporates a supercardioid directional characteristic, reducing side pickup to 12% or less than half that of conventional cardioids.

Another problem, particularly with powerful sound reinforcement systems, is mechanical (handling) noise. Aside from disturbing the audience, it can actually damage equipment. As can be seen in the cutaway, the MD431 is actually a microphone within a microphone. The dynamic transducer element is mounted within an inner capsule, isolated from the outer housing by means of a shock absorber. This protects it from handling noise as well as other mechanical vibrations normally encountered in live performances.

To screen out noise still further, an internal electrical high pass filter network is incorporated to insure that low-frequency disturbances will not affect the audio signal. A built-in mesh filter in front of the diaphragm reduces the popping and excessive sibilance often produced by close micing.

The microphone case is a heavy-duty cast outer housing with a stainless steel front grille and reed-type on-off switch. A hum bucking coil is mounted behind the transducer to cancel out any stray magnetic fields.

FIGURE 1.85   Shure SM58 vocal microphone. Courtesy Shure Incorporated.

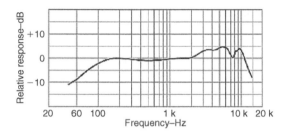

FIGURE 1.86   Frequency response of the Shure SM58 vocal microphone. Courtesy Shure Incorporated.

TABLE 1.2 Suggested Placement for the SM58 Microphone

Application	Suggested Microphone Placement	Tone Quality
Lead and backup vocals	Lips less than 150 mm (6 in) away or touching the windscreen, on axis to microphone	Robust sound, emphasized bass, maximum isolation from other sources
Speech	from mouth, just above nose height	Natural sound, reduced bass
	200 mm (8 in) to 0.6 m (2 ft) away from mouth, slightly off to one side	Natural sound, reduced bass, and minimal "s" sounds
	1 m (3 ft) to 2 m (6 ft) away	Thinner; distant sound; ambience

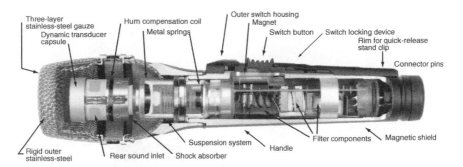

Three-layer
stainless-steel gauze
Dynamic transducer
capsule

Hum compensation coil

Metal springs

Outer switch housing
Magnet

Switch button   Switch locking device

Rim for quick-release
stand clip

Connector pins

Magnetic shield

Filter components

Handle

Suspension system

Shock absorber

Rear sound inlet

Rigid outer
stainless-steel

FIGURE 1.87   Cutaway view of a Sennheiser MD431 handheld entertainment microphone. Courtesy Sennheiser Electronic Corporation.

1.6.8 Pressure-Gradient Condenser Microphones

One of the most popular studio microphones is the Neumann U-87 multidirectional condenser microphone, Fig. 1.88, and its cousin, the Neumann U-89, Fig. 1.89. This microphone is used for close micing where high SPLs are

FIGURE 1.88   Neumann U-87 microphone. Courtesy Neumann USA.

FIGURE 1.89　Neumann U-89 microphone. Courtesy Neumann USA.

commonly encountered. The response below 30 Hz is rolled off to prevent low-frequency blocking and can be switched to 200 Hz to allow compensation for the bass rise common to all directional microphones at close range.

The figure 8 characteristic is produced by means of two closely spaced or assembled cardioid characteristic capsules, whose principal axes are pointed in opposite directions and are electrically connected in antiphase.

These microphones are usually made with backplates equipped with holes, slots, and chambers forming delay elements whose perforations act as part friction resistances and part energy storage (acoustic inductances and capacitances), giving the backplate the character of an acoustic low-pass network. In the cutoff range of this low-pass network, above the transition frequency f_t, the membrane is impinged upon only from the front, and the microphone capsule changes to a pressure or interference transducer.

The output voltage $e(t)$ of a condenser microphone using dc polarization is proportional to the applied dc voltage E_o and, for small displacement amplitudes of the diaphragm, to the relative variation in capacity $c(t)/C_o$ caused by the sound pressure

$$e(t) = E_o \frac{c(t)}{C_o} \qquad (1.25)$$

where,
E_o is the applied dc voltage,
$c(t)$ is the variable component of capsule capacity,
C_o is the capsule capacity in the absence of sound pressure,
t is the time.

The dependence of output voltage $e(t)$ on E_o is utilized in some types of microphones to control the directional characteristic. Two capsules with cardioid characteristics as shown in Fig. 1.90 are placed back to back. They can also be assembled as a unit with a common backplate. The audio (ac) signals provided by the two diaphragms are connected in parallel through a capacitor C. The intensity and phase relationship of the outputs from the two capsule halves can be affected by varying the dc voltage applied to one of them (the left cartridge in Fig. 1.90). This can be accomplished through a switch, or a potentiometer. The directional characteristic of the microphone may be changed by remote control via long extension cables.

If the switch is in its center position C, then the left capsule-half does not contribute any voltage, and the microphone has the cardioid characteristic of the right capsule-half. In switch position A, the two ac voltages are in parallel, resulting in an omnidirectional pattern. In position E the two halves are connected in antiphase, and the result is a figure 8 directional response pattern.

The letters A to E given for the switch positions in Fig. 1.90 produce the patterns given the same letters in Fig. 1.91.

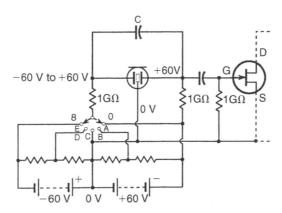

FIGURE 1.90   Circuit of the Neumann U-87 condenser microphone with electrically switchable direction characteristic. Courtesy Neumann USA.

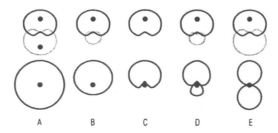

FIGURE 1.91   By using a microphone as shown in Fig. 1.89 and superimposing two cardioid patterns (top row), directional response patterns (bottom row) can be obtained. Courtesy Neumann USA.

1.6.9 Interference Tube Microphones

The *interference tube microphone*[8] as described by Olson in 1938 is often called a *shotgun microphone* because of its physical shape and directional characteristics.

Important characteristics of any microphone are its sensitivity and directional qualities. Assuming a constant sound pressure source, increasing the microphone to the source distance requires an increase in the gain of the amplifying system to produce the same output level. This is accompanied by a decrease in SNR and an increase in environmental noises including reverberation and background noise to where the indirect sound may equal the direct sound. The wanted signal then deteriorates to where it is unusable. Distance limitations can be overcome by increasing the sensitivity of the microphone, and the effect of reverberation and noise pickup can be lessened by increasing the directivity of the pattern. The interference tube microphone has these two desirable qualities.

The DPA 4017 is a supercardioid shotgun microphone. It is 8.3 inches (210 mm) long and weighs 2.5 oz (71 g), making it useful on booms, Fig. 1.92. The polar pattern is shown in Fig. 1.93.

The difference between interference tube microphones and standard microphones lies in the method of pickup.

An interference tube is mounted over the diaphragm and is schematically drawn in Fig. 1.94.

The microphone consists of four parts as shown in the schematic:

1. Interference tube with one frontal and several lateral inlets covered with fabric or other damping material.
2. Capsule with the diaphragm and counter electrode(s).
3. Rear inlet.
4. Electronic circuit.

The directional characteristics are based on two different principles:

1. In the low-frequency range, tube microphones behave as first-order directional receivers. The tube in front of the capsule can be considered as an acoustic element with a compliance due to the enclosed air volume and a

FIGURE 1.92   A supercardioid interference tube DPA 4017 microphone. Courtesy DPA Microphones, Inc.

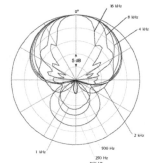

A. Directional characteristics.

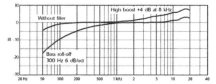

B. Frequency response of switching filters.

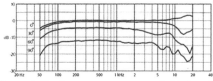

C. On and off axis response measured at 60 cm (23.6 in).

FIGURE 1.93   Directional characteristics and frequency response of a DPA 4017 microphone. Courtesy DPA Microphones, Inc.

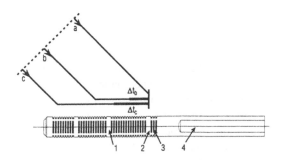

FIGURE 1.94   Schematic of an interference tube microphone.

resistance determined by the lateral holes or slits of the tube. The rear inlet is designed as an acoustic low-pass filter to achieve the phase shift for the desired polar pattern (normally cardioid or supercardioid).

2. In the high-frequency range, the acoustical properties of the interference tube determine the polar patterns. The transition frequency between the two different directional characteristics depends on the length of the tube and is given by

$$f_o = \frac{c}{2L} \qquad (1.26)$$

where,

f_o is the transition frequency,

c is the velocity of sound in air in feet per second or meters per second,

L is the length of the interference tube in feet or meters.

Referring to Fig. 1.94, if the tube is exposed to a planar sound wave, every lateral inlet is the starting point of a new wave traveling inside the tube toward the capsule as well as toward the frontal inlet. Apart from the case of frontal sound incidence, every particular wave covers a different distance to the capsule and, therefore, arrives at a different time. Figure 1.94 shows the delay times of waves b and c compared to wave a. Note that they increase with the angle of sound incidence.

The resulting pressure at the capsule can be calculated by the sum of all particular waves generated over the tube's length, all with equal amplitudes but different phase shifts. The frequency and phase response curves can be described by

$$\frac{P\theta}{P\,(\theta = 0°)} = \frac{\sin\left[\dfrac{\pi L}{\lambda} \times (1 - \cos\theta)\right]}{\dfrac{\pi L}{\lambda} \times (1 - \cos\theta)} \qquad (1.27)$$

where,

$P(\theta)$ is the microphone output at a given angle of sound incidence,

$P(\theta = 0°)$ is the microphone output along principal axis,

λ is the wavelength,

L is the length of the tube,

θ is the angle of sound incidence.

The calculated curves and polar patterns are plotted in Figs. 1.95 and 1.96 for a tube length of 9.8 inches (25 cm) without regard to the low-frequency directivity caused by the rear inlet. The shape of the response curves looks similar to that of a comb filter with equidistant minima and maxima decreasing with 6 dB/octave. The phase response is frequency independent only for frontal sound incidence. For other incidence angles, the phase depends linearly on frequency, so that the resulting pressure at the capsule shows an increasing delay time with an increasing incidence angle.

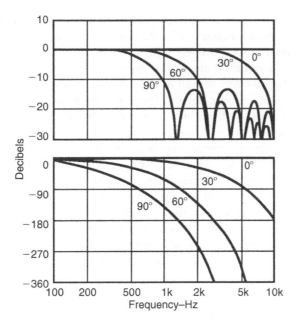

FIGURE 1.95 Calculated frequency and phase response curves of an interference tube microphone (250 mm) without rear inlet for different angles of sound incidence. Courtesy Sennheiser Electronic Corporation.

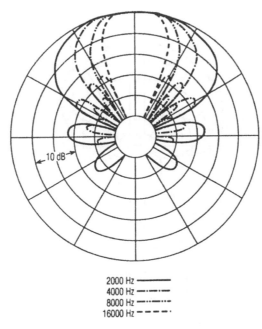

FIGURE 1.96 Calculated polar patterns of an interference tube microphone (250 mm) without rear inlet. Courtesy Sennheiser Electronic Corporation.

In practice, interference tube microphones show deviations from this simplified theoretical model. Figure 1.97 is the polar pattern of the Sennheiser MKH 60P48. The built-in tube delivers a high-frequency roll-off for lateral sound incidence with a sufficient attenuation especially for the first side lobes. The shape of the lateral inlets as well as the covering material influences the frequency and phase response curves. The transition frequency can be lowered with an acoustic mass in the frontal inlet of the tube to increase the delay times for low frequencies.

Another interference tube microphone is the Shure SM89, Fig. 1.98.[9] In this microphone, a tapered acoustic resistance is placed over the elongated interference tube slit, varying the effective length of the tube with frequency so that L/M (the ratio of tube length to wavelength) remains nearly constant over the desirable frequency range. This allows the polar response to be more consistent as frequency increases, Fig. 1.99, because the resistance in conjunction with the compliance of the air inside the tube forms an acoustical low-pass filter. High frequencies are attenuated at the end of the tube because it is the high-resistance end, allowing the high frequencies to enter the tube only near the diaphragm. This makes the tube look shorter at high frequencies, Eq. 1.27.

While a cardioid microphone may be capable of picking up satisfactorily at 3 ft (1 m), a cardioid in-line may reach 6–9 ft (1.8–2.7 m), and a super in-line may reach as far as 40 ft (12 m) and be used for picking up a group of persons in a crowd from the roof of a nearby building, following a horse around a race

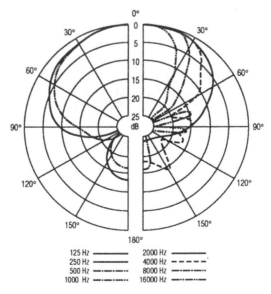

FIGURE 1.97   Characteristic of a supercardioid microphone (MKH 60P48). Courtesy Sennheiser Electronic Corporation.

track, picking up a band in a parade, and picking up other hard-to-get sounds from a distance.

There are precautions that should be followed when using interference tube microphones. Because they obtain directivity by cancellation, frequency response and phase are not as smooth as omnidirectional microphones. Also, since low frequencies become omnidirectional, the frequency response drops rapidly below 200 Hz, which helps control directivity.

It should not be assumed that no sound will be picked up outside the pickup cone. As the microphone is rotated from an on-axis position to a 180° off-axis position, there will be a progressive drop in level. Sounds originating at angles of 90° to 180° off-axis will cancel by 20 dB or more; however, the amount of cancellation depends on the level and distance of the microphone from the sound source. As an example, if an on-axis sound originated at a distance of 20 ft (6 m), a 90° to 180° off-axis sound occurring at the same distance and intensity will be reduced by 20 dB or more, providing none of the off-axis sound

FIGURE 1.98   Shure SM89 condenser shotgun microphone. Courtesy Shure Incorporated.

is reflected into the front of the microphone by walls, ceiling, and so on. On the other hand, should the off-axis sound originate at a distance of 2 ft (0.6 m) and at the same sound pressure level as the sound at 20 ft (6 m) on axis, it will be reproduced at the same level. The reason for this behavior is that the microphone is still canceling the unwanted sound as much as 20 dB, but due to the difference in the distances of the two sounds, the off-axis sound is 20 dB louder than the on-axis sound at the microphone. Therefore, they are reproduced at the same level. For a pickup in an area where random noise and reverberation are problems, the microphone should be located with the back end to the source of unwanted sound and as far from the disturbances as possible.

If the microphone is being used inside a truck or enclosed area, and pointing out a rear door, poor pickup may be experienced because all sounds, both wanted and unwanted, arrive at the microphone on-axis. Since the only entrance is through the truck door, no cancellation occurs because the truck walls inhibit the sound from entering the sides of the microphone. In this instance, the microphone will be operating as an omnidirectional microphone. Due to the reflected sound from the walls, the same condition will prevail in a room where the microphone is pointed through a window or when operating in a long hallway. For good pickup, the microphone should be operated in the open and not in closely confined quarters.

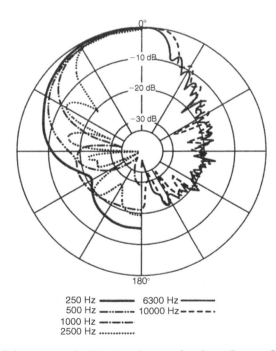

250 Hz ——————   6300 Hz ——————
500 Hz —··—··—··—   10000 Hz — — — ·
1000 Hz —·—·—·—
2500 Hz ··············

FIGURE 1.99   Polar response of a 0.5 m long shotgun microphone. Courtesy Shure Incorporated.

A shotgun interference tube microphone cannot be compared to a zoom lens since the focus does not vary nor does it reach out to gather in the sound. What the narrow polar pattern and high rate of cancellation do are to reduce pickup of the random sound energy and permit the raising of the amplifier gain following the microphone without seriously decreasing the SNR.

Difficulties may also be encountered using interference tube microphones on stage and picking out a talker in the audience, particularly where the voice is 75–100 ft (23–30 m) away and fed back through a reinforcement system for the audience to hear. Under these circumstances, only about 30–50 ft (9–15 m) is possible without acoustic feedback; even then, the system must be balanced very carefully.

1.6.10 Rifle Microphones

The *rifle microphone* consists of a series of tubes of varied length mounted in front of either a capacitor or dynamic transducer diaphragm, Fig. 1.100. The transducer may be either a capacitor or dynamic type. The tubes are cut in lengths from 2–60 inches (5–150 m) and bound together. The bundling of the tubes in front of the transducer diaphragm creates a distributed sound entrance, and the omnidirectional transducer becomes highly directional.

Sound originating on the axis of the tubes first enters the longest tube and, as the wave front advances, enters successively shorter tubes in normal progression until the diaphragm is reached. Sounds reaching the diaphragm from the source travel the same distance, regardless of the tube entered, so all sounds arriving on-axis are in phase when they reach the diaphragm. Sounds originating 90° off-axis enter all tubes simultaneously. A sound entering a longer tube may travel 18 inches (46 cm) to reach the diaphragm, while the same sound traveling through the shortest tube will travel only 3 inches (7.6 cm), with other differences for the varied length of tubing causing an out-of-phase signal at the

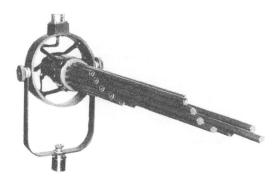

FIGURE 1.100   RCA rifle microphone. Courtesy of Radio Corporation of America.

diaphragm. Under these conditions, a large portion of the sound originating at 90° is canceled, and from 180° an even greater phase difference occurs, and cancellation is increased considerably.

The RCA MI-100006A varidirectional microphone, Fig. 1.100, consists of nineteen 5/16 inches (0.8 cm) plastic tubes, ranging from 3–18 inches (7.6–46 cm) in length. The tubes are bundled and mounted in front of an omnidirectional capacitor-microphone head. Rifle microphones are not used very much today.

1.6.11 Parabolic Microphones

Parabolic microphones use a parabolic reflector with a microphone to obtain a highly directional pickup response. The microphone diaphragm is mounted at the focal point of the reflector, Fig. 1.101. Any sound arriving from an angle other than straight on will be scattered and therefore will not focus on the pickup. The microphone is focused by moving the diaphragm in or out from the reflector for maximum pickup. This type of concentrator is often used to pickup a horse race or a group of people in a crowd.

The greatest gain in sound pressure is obtained when the reflector is large compared to the wavelength of the incident sound. With the microphone in focus, the gain is the greatest at the mid-frequency range. The loss of high frequencies may be improved somewhat by defocusing the microphone a slight amount, which also tends to broaden the sharp directional characteristics at the higher frequencies. A bowl 3 ft (0.91 m) in diameter is practically nondirectional below 200 Hz but is very sharp at 8000 Hz, Fig. 1.102. For a diameter of 3 ft, the gain over the microphone without the bowl is about 10 dB and, for a 6 ft (1.8 m) diameter bowl, approximately 16 dB.

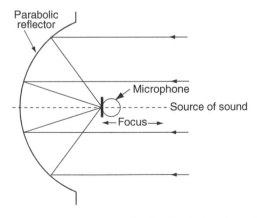

FIGURE 1.101   A parabolic bowl concentrator for directional microphone pickup.

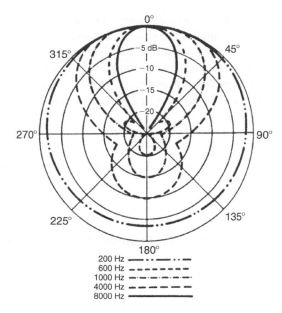

FIGURE 1.102   Polar pattern for a parabolic concentrator.

1.6.12 Zoom Microphones

A *zoom microphone*,[10] or *variable-directivity microphone*, is one that operates like and in conjunction with a zoom lens. This type of microphone is useful with television and motion-picture operations.

The optical perception of distance to the object is simply determined by the shot angle of the picture. On the other hand, a sound image is perceived by:

- Loudness.
- Reverberation (ratio of direct sound to reflected sound).
- Acquired response to sound.
- Level and arriving time difference between the two ears.

If the sound is recorded in monophonic, the following factors can be skillfully combined to reproduce a natural sound image with respect to the perceived distance:

- *Loudness:* Perceived loudness can be controlled by varying microphone sensitivity.
- *Reverberation:* The representation of the distance is made by changing the microphone directivity or the ratio between direct and reverberant sound. In a normal environment, we hear a combination of direct sound and its reflections. The nearer a listening point is to the source, the larger

the ratio of direct to reverberant sound. The farther the listening point is from the source, the smaller the ratio; therefore, use of a high-directivity microphone to keep direct sound greater than reflected sound permits the microphone to get apparently closer to the source by reducing reverberant sound pickup. For outdoor environments, use of directional microphones allows the ambient noise level to be changed for natural representation of distances.

- *Acquired human response to sound:* Normally we can tell approximately how far a familiar object as a car or a person is by the sound generated by the objects because we acquire the response to sound through our daily experiences.

1.6.12.1 Distance Factor

The fact that microphone directivity determines the perceived distance can be explained from the viewpoint of the distance factor. Figure 1.103 shows the sound pressure level at the position of an omnidirectional microphone versus the distance between the microphone and a sound source S, with ambient noise evenly distributed. Suppose the distance is 23 ft (7 m) and the ambient noise level is at 1. If the microphone is replaced by one that has a narrow directivity with the same on-axis sensitivity, less noise is picked up, so the observed noise level is lowered to 2. For an omnidirectional microphone, the same effect can be obtained at a distance of 7.5 ft (2.3 m). From a different standpoint, the same SNR as for an omnidirectional microphone at 20.6 ft (6.3 m) can be obtained at a distance of 65 ft (20 m). The ratio of actual-to-observed distance is called the *distance factor*.

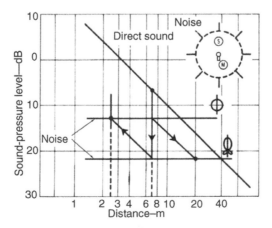

FIGURE 1.103   Relationship between sound pressure level and distance in an evenly distributed noise environment.

1.6.12.2 Operation of Zoom Microphones

By changing the sensitivity and directivity of a microphone simultaneously, an acoustical zoom effect is realized, and more reality becomes possible in sound recording. Figure 1.104 is the basic block diagram of a zoom microphone system. The system consists of three unidirectional microphone capsules (1 through 3) arranged on the same axis. The three capsules have the same characteristics, and capsule 3 faces the opposite direction. The directivity can be varied from omnidirectional to second-order gradient unidirectional by varying the mixing ratio of the output of each capsule and changing the equalization characteristic accordingly. An omnidirectional pattern is obtained by simply combining the outputs of capsule 2 and 3. In the process of directivity change from omnidirectional to unidirectional, the output of capsule 3 is gradually faded out, while the output of capsule 1 is kept off. Furthermore, the equalization characteristic is kept flat, because the on-axis frequency response does not change during this process. In the process of changing from unidirectional to second-order

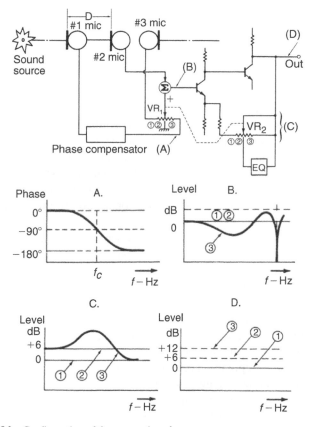

FIGURE 1.104   Configuration of the zoom microphone.

gradient unidirectional, the output of capsule 3 is kept off. The second-order gradient unidirectional pattern is obtained by subtracting the output of capsule 1 from the output of capsule 2. To obtain the second-order gradient unidirectional pattern with minimum error, the output level of capsule 1 needs to be trimmed. Since the on-axis response varies according to the mixing ratio, the equalization characteristics also have to be adjusted along with the level adjustment of the output of capsule 1. The on-axis sensitivity increase of second-order gradient setup over the unidirectional setup allows the gain of the amplification to be unchanged.

1.6.12.3 Zoom Microphone Video Camera Linkage

In order to obtain a good matching of picture and sound, a mechanism that synchronizes the optical zooming and acoustical zooming becomes inevitable. Electrical synchronization would also be possible by using voltage-controlled amplifiers (VCA) or voltage-controlled resistors (VCR).

1.6.13 Automatic Microphone Systems

There have been many advances in automatic mixers where the microphone is normally off until gated on by a signal, hopefully, a wanted signal. Many operate on an increased level in one or more microphones with respect to the random background noise.

While the *Shure Automatic Microphone System* (AMS) is a discontinued microphone, it is still used in many venues. The system turns microphones on and off (with automatic gating), greatly reducing the reverberant sound quality and feedback problems often associated with the use of multiple microphones. The AMS microphones are gated on only by sounds arriving from the front within their acceptance angle of 120°. Other sounds outside the 120° angle, including background noise, will not gate the microphones on, regardless of level. In addition, the AMS adjusts gain automatically to prevent feedback as the number of microphones increases.

The Shure Model AMS22 low-profile condenser microphone, Fig. 1.105, is designed for use only with the Shure AMS. Unlike conventional microphones, it contains electronic circuitry and a novel transducer configuration to make it compatible with the Shure AMS mixers. The microphone should not be connected to standard simplex- (phantom-) or non-simplex powered microphone inputs because they will not function properly.

AMS microphones, in conjunction with the special circuitry in the AMS mixers, uniquely discriminate between desired sounds that originate within their 120° front acceptance angle and all other sounds. Sounds from the front of a microphone are detected and cause it to be gated on, transmitting its signal to the mixer output. Sounds outside the acceptance angle will not gate the microphone on. When an AMS22 is gated on, it operates like a hemi- or half-cardioid microphone because half the cardioid pattern

FIGURE 1.105   Shure Automatic Microphone System (AMS) model AMS22 low-profile microphone. Courtesy Shure Incorporated.

disappears when the microphone is placed on a surface, Fig. 1.106. Each AMS microphone operates completely independently in analyzing its own sound field and deciding whether or not a sound source is within the front acceptance angle.

The microphone should be placed so that intended sources are within 60° of either side of the front of the microphone—that is, within 120° acceptance angle. Sources of undesired sound should be located outside the 120° acceptance angle. Each microphone should be at least 3 ft from the wall behind it, and items such as large ashtrays or briefcases should be at least 1 ft behind it. If the microphones are closer than that, reflections will reduce the front-to-back discrimination and, therefore, make the microphone act more like a conventional cardioid type.

1.6.14 PolarFlex™ Microphone System

The PolarFlex™ system by Schoeps models any microphone. The system features two output channels with two microphones per channel, Fig. 1.107. The standard system consists of an omnidirectional and a figure 8 microphone for each channel and an analog/digital processor.

Essential sonic differences between condenser microphones of the same nominal directional pattern are not only due to frequency response, but also to the fact that the polar pattern is not always uniformly maintained throughout the entire frequency range particularly at the lowest and highest frequencies. Though ostensibly a defect, this fact can also be used to advantage (e.g., adaptation to the acoustic of the recording room). While the frequency response

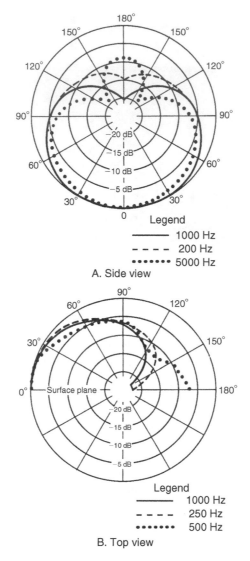

FIGURE 1.106 Hemicardioid polar pickup pattern for a Shure AMS surface microphone. Courtesy Shure Incorporated.

at a given pickup angle can be controlled by equalizers, there was no way to alter the polar pattern correspondingly. The only way to control this situation was through the choice of microphones having different variations of the polar pattern versus frequency. With the DSP-4P processor, nearly ideal directional characteristics can be selected, and nearly any frequency-dependent directional characteristic that may be desired—e.g., a cardioid becomes

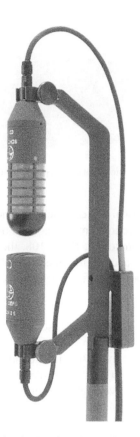

FIGURE 1.107   A Schoeps PolarFlex™ microphone with an omnidirectional and a figure 8 microphone. Courtesy Schoeps GmbH.

omnidirectional below the midrange, so that it has better response at the very lowest frequencies. Also modeling a large-diaphragm microphone is possible.

Furthermore, in excessively reverberant spaces one could record a drier sound (cardioid or supercardioid setting) or, in spaces that are dry, accept more room reflections (wide cardioid or omni setting) in the corresponding frequency range.

In such cases it is not the frequency response but rather the ratio of direct to reflected sound, that will be altered. That cannot be done with an equalizer nor can a reverb unit reduce the degree of reflected sound after the fact.

In the arrangement of Fig. 1.107, an omnidirectional microphone with a mild high-frequency emphasis in the direct sound field is used. Because of its angle of orientation, the capsule has ideal directional response in the horizontal plane; the high-frequency emphasis compensates for the high-frequency losses due to lateral sound incidence.

A figure 8 microphone is set directly above the omni. The direction to which it is aimed will determine the orientation of the resulting adjustable virtual

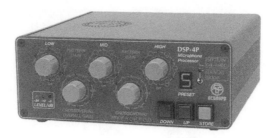

FIGURE 1.108   Schoeps DSP-4P microphone processor. Courtesy Schoeps GmbH.

microphone. The hemispherical device attached to the top of the figure 8 flattens the response of the omnidirectional microphone at the highest frequencies.

By using the DSP-4P processor, Fig. 1.108, the following settings can be made independently of one another in three adjustable frequency ranges. With the three knobs in the upper row, the directional patterns in each of the three frequency bands can be set. The settings are indicated by a circle of LEDs around each of the knobs. At the lower left of each knob is the omnidirectional setting; at the lower right is the figure 8 setting. Eleven intermediate pattern settings are available. The knobs in the lower row are set between those in the upper row. They are used for setting the boundaries between the frequency ranges, 100 Hz–1 kHz and 1–10 kHz, respectively, in ⅓ octave steps.

The three buttons at the lower right are for storing and recalling presets. If the unprocessed microphone signals have been recorded, these adjustments can be made during post processing.

The processor operates at 24 bit resolution with either a 44.1 kHz or 48 kHz sampling rate. When a digital device is connected to the input, the PolarFlex™ processor adapts to its clock signal.

1.7 STEREO MICROPHONES

Stereo microphones are microphones or systems used for coincident, XY, M/S, SASS, binaural in-the-head, and binaural in-the-ear (ITE) recording. These systems have the microphones close together (in proximity of a point source or ear-to-ear distance) and produce the stereophonic effect by intensity stereo, time-based stereo, or a combination of both.

1.7.1 Coincident Microphones

A highly versatile stereo pickup is the *coincident microphone technique*.[11,12,13] Coincident means that sound reaches both microphones at the same time, implying that they are at the same point in space. In practice, the two microphones cannot occupy the same point, but they are placed as closely together as possible. There are special-purpose stereo microphones available that combine the two microphones in one case. Since they are essentially at the same point, there can be no time differences between arrival of any sound from any direction; thus no cancellation can occur. It might first appear that there could be no stereophonic result from this configuration. The two microphones are usually unidirectional and oriented at 90° to one another. The combination is then aimed at the sound source, each microphone 45° to a line through the source. Stereo results from intensity differences—the left microphone (which is to the right of the pair) will receive sounds from the left-hand part of the stage with greater volume than it will receive from the right-hand side of the stage.

The stereo result, although often not as spectacular as that obtained from spaced microphones, is fully mono compatible, and it most accurately reproduces the sound of the acoustic environment. It is quite foolproof and quick to set up.

Variations of the coincident technique include changing the angle between the microphone (some stereo microphones are adjustable); using bidirectional microphones, which results in more reverberant sound; using combinations of unidirectional and bidirectional microphones; and using matrix systems, which electrically provide sum and difference signals from the left and right channels (these can be manipulated later for the desired effect).

The basic coincident technique was developed in the 1930s (along with the first stereo recordings) by English engineer Alan Blumlein.[14] Blumlein used two figure 8 pattern ribbon microphones mounted so that their pattern lobes were at right angles (90°) to each other, as shown in Fig. 1.109. The stereo effect is produced primarily by the difference in amplitude generated in the two microphones by the sound source. A sound on the right generates a larger signal in microphone B than in microphone A. A sound directly in front produces an equal signal in both microphones, and a sound on the left produces a larger signal in microphone A than in microphone B. The same process takes place with spaced omnidirectional microphones, but because of the spacing, there is

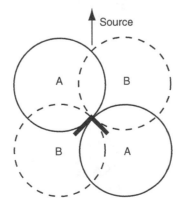

FIGURE 1.109   Coincident microphone technique using two bidirectional microphones.

also a time delay between two signals (comb filter effect). It can also produce a loss in gain and unpleasant sound if the two channels are combined into a single mono signal. Since the coincident microphone has both its transducers mounted on the same vertical axis, the arrival time is identical in both channels, reducing this problem to a large degree.

Modern coincident microphones often use cardioid or hypercardioid patterns. These patterns work as well as the figure 8 pattern microphones in producing a stereo image, but they pickup less of the ambient hall sound.

Probably the strongest virtue of the coincident microphone technique is its simplicity under actual working conditions. Just place the microphone in a central location that gives a good balance between the musicians and the acoustics of the hall. It is this simplicity that makes coincident microphones a favorite of broadcast engineers recording (or transmitting) live symphonic concerts.

1.7.2 XY Stereo Technique

The XY technique uses two identical directional microphones that, in relation to the recording axis, are arranged at equal and opposed offset angles. The leftward pointing X microphone supplies the L signal directly, and the rightward pointing Y microphone supplies the R signal, Fig. 1.110. The stereophonic properties depend on the directional characteristics of the microphones and the offset angle.

One property specific to a microphone system is the recording angle, which defines the angle between the center axis (symmetry axis of the system) and the direction where the level differences between the L and R define the angular

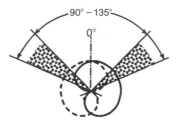

A. Cardioid microphones

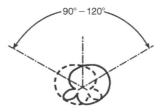

B. Supercardioid microphones

FIGURE 1.110   XY stereo technique patterns.

range of sound incidence where regular stereophonic reproduction is obtained. In most cases there is another opening for backward sound reception besides the recording angle for frontal sound pickup.

Another important aspect concerns the relationship between the sound incidence angle and the stereophonic reproduction angle. As both XY and M/S recording techniques supply pure intensity cues, a relationship can be applied that relates the reproduction angle to the level difference of the L and R signals for the standard listening configuration based on an equilateral triangle, Fig. 1.111. This relationship is shown in Fig. 1.112 and is valid at frequencies between 330 and 7800 Hz within ±3°. The level difference is plotted on the horizontal axis and the reproduction angle can be read on the vertical scale. A 0° reproduction angle means localization at the center of the stereo base, and 30° means localization at one of the loudspeakers.

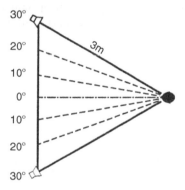

FIGURE 1.111   Standard listening configuration.

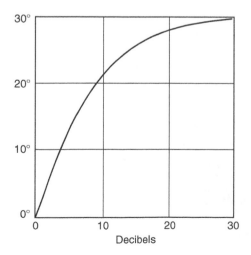

FIGURE 1.112   Stereophonic localization.

Figure 1.113 shows the XY properties of wide-angle cardioids. The lower graph illustrates that the stereo image does not cover the full base width but is rather limited to some 20° at best. The recording angle can be altered between 90° and 120°. In-phase reproduction with correct side direction is maintained for all angles of sound incidence. The downward bend of the curves indicates that the stereo image is affected by geometric compression effects.

Figure 1.114 shows the XY properties of cardioids. The recording angle can be altered between 90° and 180°. Again, in-phase reproduction at the correct side is maintained for all directions of sound incidence. As all reproduction curves touch the upper edge of the graph, full stereo width is available at all offset angles. The individual curvatures indicate deviations from the ideal geometrical reproduction performance which would be represented by a straight line. Downward bends indicate image compression at the base edges, whereas upward bends indicate expansion. The lower graph shows that compression effects occur at frontal sound pickup for offset angles above 30°, whereas expansion occurs at smaller offset angles. Best reproduction linearity is performed at offset angles around 30°. The reproduction of back sound is always affected by angular compression. Extreme compression effects occur where the curves touch the upper edge of the graph. The reproduction is then clustered at one of the loudspeakers.

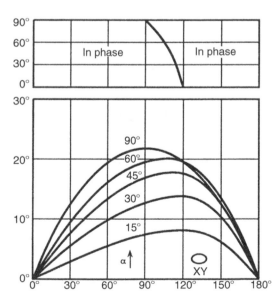

FIGURE 1.113 XY properties of wide-angle cardioids.

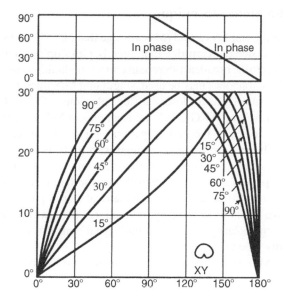

FIGURE 1.114   XY properties of cardioids.

1.7.3 The ORTF Technique

A variation on the basic XY coincident technique is the *ORTF technique*. The initials ORTF stand for *Office de Radiodiffusion Télévision Française*, the French government radio network that developed this technique. The ORTF method uses two cardioid microphones spaced 7 inches (17 cm) apart and facing outward with an angle of 110° between them, Fig. 1.115. Because of the spacing between the transducers, the ORTF method does not have the time-coherence properties of M/S or XY micing.

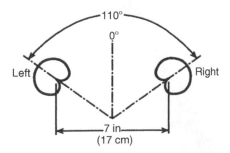

FIGURE 1.115   ORTF microphone technique.

1.7.4 The M/S Stereo Technique

The M/S technique employs a mid (M) cartridge that directly picks up the mono sum signal, and a side (S) cartridge that directly picks up the stereo difference signal (analogous to the broadcast stereo subcarrier modulation signal). Although two individual microphones may be used, single-unit M/S microphones are more convenient and generally have closer cartridge placement. Figure 1.116, a Shure VP88, and Fig. 1.117, an AKG C422, are examples of M/S microphones.

Figure 1.118 indicates the pickup patterns for a typical M/S microphone configuration. The mid cartridge is oriented with its front (the point of greatest sensitivity) aimed at the center of the incoming sound stage. A cardioid (unidirectional) pattern as shown is often chosen for the mid cartridge, although other patterns may also be used. For symmetrical stereo pickup, the side cartridge must have a side-to-side facing bidirectional pattern (by convention, the lobe with the same polarity as the front mid signal aims 90° to the left, and the opposite polarity lobe to the right).

In a stereo FM or television receiver, the mono sum baseband signal and the stereo difference subcarrier signal are demodulated and then decoded, using a

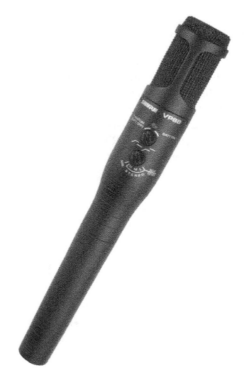

FIGURE 1.116   Shure VP88 stereo condenser microphone. Courtesy Shure Incorporated.

FIGURE 1.117   AKG C422 stereo coincident microphone. Courtesy AKG Acoustics, Inc.

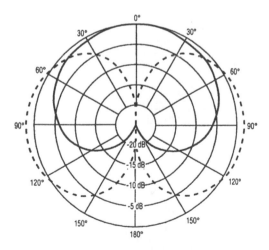

FIGURE 1.118   MS microphone pickup patterns.

sum-and-difference matrix, into left and right stereo audio signals. Similarly, the mid (mono) signal and the side (stereo difference) signal of the MS microphone may be decoded into useful left and right stereo signals.

The mid cartridge signal's relation to the mono sum signal, and the side cartridge signal's relation to the stereo difference signal, can be expressed simply by

$$M = \frac{1}{2(L + R)} \tag{1.28}$$

$$S = \frac{1}{2(L - R)} \tag{1.29}$$

Solving for the left and right signals,

$$L = M + S \tag{1.30}$$

$$R = M - S \tag{1.31}$$

Therefore, the left and right stereo signals result from the sum and difference, respectively, of the mid and side signals. These stereo signals can be obtained by processing the mid and side signals through a sum and difference matrix, implemented with transformers, Fig. 1.119, or active circuitry. This matrix may be external to the M/S microphone or built in.

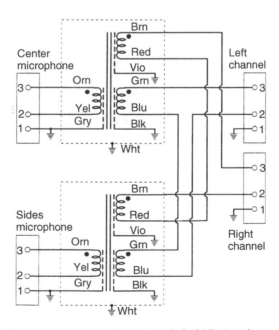

FIGURE 1.119   Transformer sum and difference matrix for M/S microphones.

In theory, any microphone pattern may be used for the mid signal pickup. Some studio M/S microphones provide a selectable mid pattern. In practice, however, the cardioid mid pattern is most often preferred in M/S microphone broadcast applications.

The AKG C422 shown in Fig. 1.120 is a studio condenser microphone that has been specially designed for sound studio and radio broadcasting. The microphone head holds two twin diaphragm condenser capsules elastically suspended to protect against handling noise.

The wire-mesh grille is differently colored at the two opposing grille sides (light is the front grille side; dark is the rear grille side), thereby allowing the relative position of the two systems to be visually checked. The entire microphone can be rotated 45° about the axis to allow quick and exact changeover from 0° for M/S to 45° (for XY stereophony) even when the microphone is rigidly mounted. The upper microphone cartridge can be rotated 180° with respect to the lower one. A scale on the housing adjustment ring and an arrow-shaped mark on the upper system allows the included angle to be exactly adjusted. In sound studio work and radio broadcasts, it is often necessary to recognize the respective positions of the two systems from great distances; therefore, two light-emitting diodes with a particularly narrow light-emitting angle are employed. One is mounted in the upper (rotatable) housing, and the other in the lower (nonrotatable) housing. To align the heads, simply have the units rotated until the light-emitting diode is brightest on the preferred axis.

Enclosed within the microphone shaft are two separate field-effect transistor preamplifiers, one for each channel. The output level of both channels may be simultaneously lowered by 10 dB or 20 dB.

The C422 is connected to an S42E remote-control unit that allows any one of nine polar patterns to be selected for each channel. Because of noiseless selection, polar pattern changeover is possible even during recording.

Each channel of the microphone incorporates two cardioid diaphragms facing 180° of each other (back to back), Fig. 1.120. Note the 12 Vdc phantom power for the electronics and the 60 Vdc phantom power for the lower transducer (1), insuring that transducer 1 is always biased on. This transducer has a positive

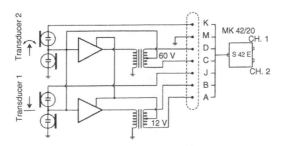

FIGURE 1.120 Schematic of the AKG C422 coincident microphone. Courtesy AKG Acoustics, Inc.

output for a positive pressure. The second or upper transducer is connected to pin K, which through the S42E, has nine switchable voltages between 0 Vdc and 120 Vdc. When the voltage at K is 60 Vdc, the output of transducer 2 is 0 (60 Vdc on either side of it), so the microphone output is cardioid.

When the voltage at K is 120 Vdc, transducer 2 is biased with 60 Vdc of an opposite polarity from transducer 1 so the output is 180° out of polarity, the mixed output being a figure 8 pattern.

When the voltage at K is 0 Vdc, transducer 2 has a 60 Vdc bias on it with the same polarity as transducer 1. Because the transducers face in opposite directions, when these two outputs are combined, an omnidirectional pattern is produced.

By varying the voltage on K between 0 Vdc and 120 Vdc, various patterns between a figure 8 and an omnidirectional pattern can be produced.

The Shure VP88 stereo microphone, Fig. 1.116, also employs a switchable pattern. Figure 1.121 shows the polar response of the mid capsule and the side capsule.

The left and right stereo signals exhibit their own equivalent pickup patterns corresponding to, respectively, left-forward-facing and right-forward-facing microphones. Figure 1.122 shows the relative levels of the mid and side microphones and the stereo pickup pattern of the Shure VP88 microphone in the L position with the bidirectional side pattern maximum sensitivity 6 dB lower than the maximum mid sensitivity. The small rear lobes of each pattern are 180° out of polarity with the main front lobes. For sound sources arriving at 0° the left and right output signals are equal, and a center image is reproduced between the loudspeakers. As the sound source is moved off-axis, an amplitude difference between left and right is created, and the loudspeaker image is moved smoothly off-center in the direction of the higher amplitude signal.

When the mid (mono) pattern is fixed as cardioid, the stereo pickup pattern can be varied by changing the side level relative to the mid level. Figure 1.123 shows an M/S pattern in the M position with the side level 1.9 dB lower than the mid level. Figure 1.124, position H, increases the side level to 1.6 dB higher than the mid level. The three resultant stereo patterns exhibit pickup angles of 180°, 127°, and 90°, respectively. The incoming sound angles, which will create left, left-center, center, right-center, and right images, are also shown. Note the changes in the direction of the stereo patterns and the size of their rear lobes.

Taking the directional properties of real microphones into consideration, it becomes clear that the M/S technique provides a higher recording fidelity than the XY technique. There are at least three reasons for this:

1. The microphones in an XY system are operated mainly at off-axis conditions, especially at larger offset angles. The influence of directivity imperfections is more serious than with MS systems, where the M microphone is aimed at the performance center. This is illustrated by Fig. 1.125.
2. The maximum sound incidence angle for the microphone is only half that of the X and Y microphones, although the covered performance area is the same for all microphones. This area is symmetrically picked up by the M microphone,

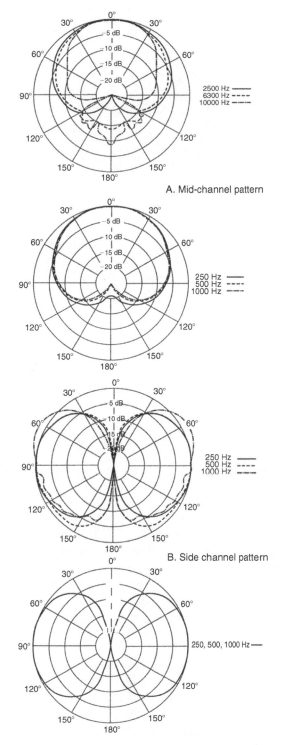

A. Mid-channel pattern

B. Side channel pattern

FIGURE 1.121 Polar response of the Shure VP88 M/S microphone. Courtesy Shure Incorporated.

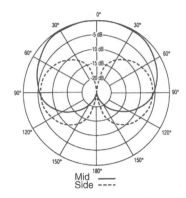

A. Relative levels of the mid and side microphones.

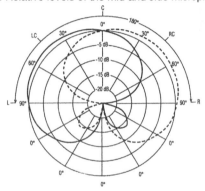

B. Pickup pattern of the system.

FIGURE 1.122   Stereo pickup pattern of the Shure VP88 in the L position. Courtesy Shure Incorporated.

but unsymmetrically by the X and Y microphones. The M/S system can supply the more accurate monophonic (M) signal in comparison with the XY system.

3. The M/S system picks up the S signal with a bidirectional microphone. The directivity performance of this type of microphone can be designed with a high degree of perfection, so errors in the S signal can be kept particularly small for all directions of sound incidence. The M/S system can supply a highly accurate side (S) signal.

4. In the M/S technique, the mono directivity does not depend on the amount of S signal applied to create the stereophonic effect. If recordings are made in the M/S format, a predictable mono signal is always captured. On the other hand, the stereophonic image can be simply influenced by modifying the S level without changing the mono signal. This can even be done during postproduction.

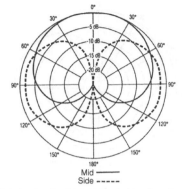

A. Relative levels of the mid and side microphones.

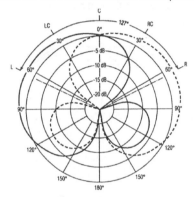

B. Pickup pattern of the system.

FIGURE 1.123   Stereo pickup pattern of the Shure VP88 in the M position. Courtesy Shure Incorporated.

1.7.5 The Stereo Boom Microphone

Microphones for stereophonic television broadcasting have their own special problems. The optimal distance from the TV screen is considered to be five times the screen diagonal, at which distance the line structure of the TV image can no longer be resolved by the human eye. The resulting minimum observer distance is therefore 11 ft (3.3 m) and the two loudspeakers should be at least 12.5 ft (3.8 m) apart. This is certainly not realistic for television viewing.

The sound engineer must take into account that the reproduction will be through loudspeakers right next to the TV screen as well as through hi-fi equipment. If, for instance, the full base width is used during the sound recording of a TV scene, an actor appearing at the right edge of the picture will be heard through the right loudspeaker of the hi-fi system and sound as if he is far to the right of the TV set. This will result in an unacceptable perception of location.

Mid ——
Side ----

A. Relative levels of the mid and side microphones.

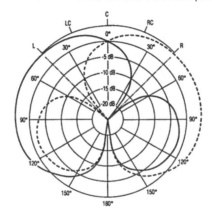

B. Pickup pattern of the system.

FIGURE 1.124   Stereo pickup pattern of the Shure VP88 in the H position. Courtesy Shure Incorporated.

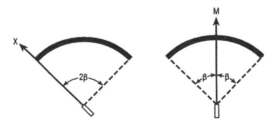

FIGURE 1.125   Exposure of X and M microphones to 0° pickup.

The viewer must be able to hear realistically the talker on the TV screen in the very place where the viewer sees the talker. To achieve this goal, German television proposed to combine a unidirectional microphone for the recording of the actors with a figure 8 microphone for the recording of the full stereophonic basic width.

This recording technique utilizes a figure 8 microphone suspended from the boom in such a way that it maintains its direction when the boom is rotated while a second microphone with unidirectional pattern is mounted on top and follows the movement of an actor or reporter, Fig. 1.126. To make sure that the directional characteristic of the moving microphone does not have too strong an influence and that a slight angular error will not lead to immediately perceptible directional changes, the lobe should be somewhat wider as with the customary shotgun microphones in use today.

It is now possible to produce a finished stereo soundtrack on location by positioning the microphones in the manner of an M/S combination. The level of the figure 8 microphone can be lowered and used only for the recording of voices outside of the picture, ambience and music. This microphone should always remain in a fixed position and its direction should not be changed. The S-signal generated in this way must be attenuated to such a degree that the M-signal microphone will always remain dominant. This microphone is the one through which the actors pictured on the screen will be heard.

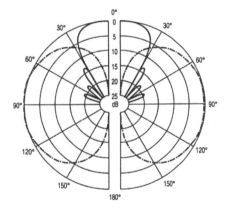

FIGURE 1.126   Stereo boom microphone using Sennheiser MKH 30 and MKH 70 microphones. Courtesy Sennheiser Electronic Corporation.

1.7.6 SASS Microphones

The Crown® SASS-P MK II or Stereo Ambient Sampling System™, Fig. 1.127, is a patented, mono-compatible, near-coincident array stereo condenser microphone using PZM technology.

The SASS uses two PZMs mounted on boundaries (with a foam barrier between them) to make each microphone directional. Another Crown model, SASS-B, is a similarly shaped stereo boundary mount for Brüel & Kjaer 4006 microphones and is used for applications requiring extremely low noise.

FIGURE 1.127 Crown® SASS-P MK II stereo microphone. Courtesy of Crown International, Inc.

Controlled polar patterns and human-head-sized spacing between capsules create a focused, natural stereo image with no hole-in-the-middle for loudspeaker reproduction, summing comfortably to mono if required.

The broad acceptance angle (125°) of each capsule picks up ambient sidewall and ceiling reflections from the room, providing natural reproduction of acoustics in good halls and ambient environments. This pattern is consistent for almost ±90° vertical.

A foam barrier/baffle between the capsules shapes the pickup angle of each capsule toward the front, limiting overlap of the two sides at higher frequencies. Although the microphone capsules are spaced a few centimeters apart, there is little phase cancellation when both channels are combined to mono because of the shadowing effect of the baffle. While there are phase differences between channels, the extreme amplitude differences between the channels caused by the baffle reduce phase cancellations in mono up to 20 kHz.

The SASS has relatively small boundaries. However, it has a flat response down to low frequencies because there is no 6 dB shelf as in standard PZM microphones (see Section 1.6.1.2.3). The flat response is attained because the capsules are omnidirectional below 500 Hz, and their outputs at low frequencies are equal in level, which, when summed in stereo listening, causes a 3 dB rise in perceived level. This effectively counteracts one half of the low-frequency shelf normally experienced with small boundaries.

In addition, when the microphone is used in a reverberant sound field, the effective low-frequency level is boosted another 3 dB because the pattern is omnidirectional at low frequencies and unidirectional at high frequencies. All of the low-frequency shelf is compensated, so the effective frequency response is uniform from 20 Hz–20 kHz. Figure 1.128 is the polar response of the left channel (the right channel is the reverse of the left channel).

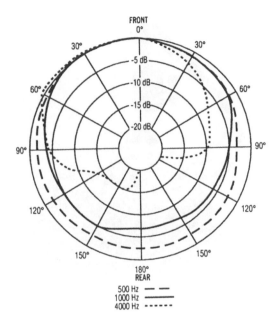

FRONT
0°

FIGURE 1.128   SASS-P MK II polar response, of the left channel. 0° sound incidence is perpendicular to the boundary. The right channel is a mirror image of the left channel. Courtesy Crown International, Inc.

1.7.7 Surround Sound Microphone System

1.7.7.1 Schoeps 5.1 Surround System

The Schoeps 5.1 surround system consists of the KFM 360 sphere microphone, two figure 8 microphones with suspension, and the DSP-4 KFM 360 processor, Fig. 1.129.

The central unit in this system is the sphere microphone KFM 360. It uses two pressure transducers and can, even without the other elements of the system, be used for stereophonic recording. Its recording angle is about 120°, permitting closer micing than a standard stereo microphone. The necessary high-frequency boost is built into the processor unit.

Surround capability is achieved through the use of two figure 8 microphones, which can be attached beneath the pressure transducers by an adjustable, detachable clamp system with bayonet-style connectors. The two microphones should be aimed forward.

The DSP-4 KFM 360 processor derives the four corner channels from the microphone signals. A center channel signal is obtained from the two front signals, using a special type of matrix. An additional channel carries only the low

FIGURE 1.129   Schoeps KFM 360 sphere microphone. Courtesy Schoeps GmbH.

frequencies, up to 70 Hz. To avoid perceiving the presence of the rear loud-speakers, it is possible to lower the level of their channels, to delay them and/or to set an upper limit on their frequency response, Fig. 1.130.

The front stereo image width is adjustable and the directional patterns of the front-facing and rear-facing pairs of virtual microphones can be chosen independently of one another.

The processor unit offers both analog and digital inputs for the microphone signals. In addition to providing gain, it offers a high-frequency emphasis for the built-in pressure transducers as well as a low-frequency boost for the figure 8s.

As with M/S recording, matrixing can be performed during postproduction in the digital domain.

The system operates as follows: the front and rear channels result from the sum (front) and difference (rear) of the omnidirectional and figure 8 microphones on each side, respectively, Fig. 1.131. The four resulting virtual microphones that this process creates will seem to be aimed forward and backward, as the figure 8s are. At higher frequencies they will seem to be aimed more toward the sides (i.e., apart). Their directional pattern can be varied, anywhere from omnidirectional to cardioid to figure 8. The pattern of the two rear-facing virtual microphones can be different from that of the two forward-facing ones. Altering the directional

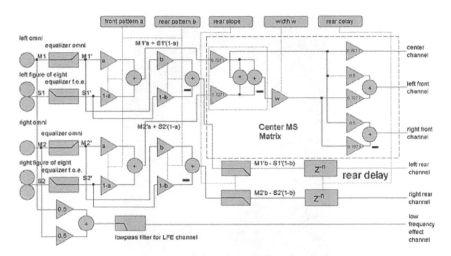

FIGURE 1.130   Schoeps DSP-4 KFM 360 processor. Courtesy Schoeps GmbH.

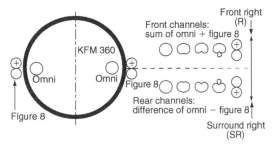

FIGURE 1.131   Derivation of the right (R) and right surround signals (SR) of the Schoeps 5.1 Surround System. Courtesy Schoeps GmbH.

patterns alters the sound as well, in ways that are not possible with ordinary equalizers. This permits a flexible means of adapting to a recording situation—to the acoustic conditions in the recording space—and this can even be done during postproduction, if the unprocessed microphones signals are recorded.

This four-channel approach yields a form or surround reproduction without a center channel—something that is not what everybody requires.

1.7.7.2 Holophone® H2-PRO Surround Sound System

The elliptical shape of the Holophone® H2-PRO emulates the characteristics of a human head, Fig. 1.132. Sound waves bend around the H2-PRO as they do around the head providing an accurate spatiality, audio imaging, and natural directionality. Capturing the directionality of these soundwaves

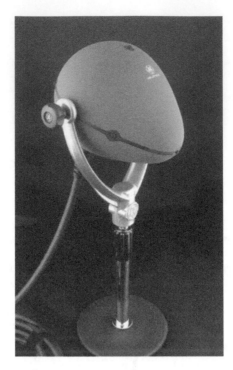

FIGURE 1.132   Holophone® H2-PRO surround sound system. Courtesy Holophone®.

translates into a very realistic surround sound experience. The total surface area of the eight individual elements combines with the spherical embodiment of the H2-PRO to capture the acoustic textures required for surround reproduction, Fig. 1.133. The embodiment acts as an acoustic lens capturing lows and clean highs.

A complete soundfield can be accurately replicated without the use of additional microphones—a simple point-and-shoot operation. The Holophone H2-PRO is capable of recording up to 7.1 channels of discrete surround sound. It terminates in eight XLR microphone cable ends (Left, Right, Center, Low Frequency, Left Surround, Right Surround, Top, and Center Rear). These co-relate to the standard 5.1 channels and add a top channel for formats such as IMAX and a center rear channel for extended surround formats such as Dolby EX, DTS, ES, and Circle Surround. Because each microphone has its own output, the engineer may choose to use as many or as few channels as the surround project requires as channel assignments are discrete all the way from the recording and mixing process to final delivery. It is well suited for television broadcasters (standard TV, DTV, and HDTV), radio broadcasters, music producers and engineers, film location recording crews, and independent project studios.

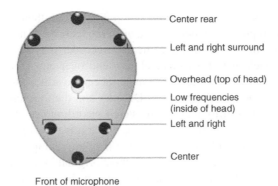

Center rear

Left and right surround

Overhead (top of head)

Low frequencies
(inside of head)

Left and right

Center

Front of microphone

FIGURE 1.133   Location of the microphones on the H2-PRO head. Courtesy Holophone®.

1.7.7.3 Holophone® H4 SuperMINI Surround Sound System

The H4 SuperMINI head, Fig. 1.134, contains six microphone elements that translate to the standard surround sound loudspeaker configuration: L, R, C, LFE, LS, RS. The LFE collects low-frequency signals for the subwoofer. The six discrete channels are fed into a Dolby® Pro-Logic II encoder which outputs the audio as a stereo signal from a stereo mini-plug to dual XLRs, dual RCAs,

FIGURE 1.134   Holophone® H4 SuperMINI Surround Sound System. Courtesy Holophone®.

or dual mini-plugs. The left and right stereo signals can then be connected to the stereo audio inputs of a video camera or stereo recorder. The encoded signal is recorded onto the media in the camera or recorder and the captured audio can be played back in full 5.1-channel surround over any Dolby® Pro-Logic II–equipped home theatre system. The material can be edited and the audio can be decoded via a Dolby® Pro-Logic II Decoder and then brought into an NLE including Final Cut or iMovie, etc. The stereo recording can also be broadcast directly through the standard infrastructure. Once it is received by a home theatre system, containing a Dolby® Pro-Logic II or any compatible decoder, the six channels are completely unfolded to their original state. Where no home theatre receiver is detected, the signal will simply be heard in stereo. The Super-MINI has additional capabilities that include an input for an external, center-channel-placed shotgun or lavalier microphone to enhance sonic opportunity options and features an audio zoom button that increases the forward bias of the pickup pattern. It also includes virtual surround monitoring via headphones for real-time on-camera 3D audio monitoring of the surround field.

1.8 MICROPHONES FOR BINAURAL RECORDING

1.8.1 Artificial Head Systems

Human hearing is capable of selecting single sounds from a mixture of sounds while suppressing the unwanted components (the cocktail party effect). This is done in the listener's brain by exploiting the ear signals as two spatially separated sound receivers in a process frequently referred to as *binaural signal processing*. A simple test will verify this statement: when listening to a recording of several simultaneous sound events recorded by a single microphone, the individual sources cannot be differentiated.

Two spaced microphones or more elegant multielement spatially sensitive microphones, such as a stereo coincident microphone, have been used to capture the spatial characteristics of sounds, but they have frequently been deficient when compared to what a person perceives in the same environment. This lack of realism is attributed to absence of the spectrum modification inherent in sound propagation around a person's head and torso and in the external ear—i.e., the transfer function of the human and the fact that the signals are kept separate until very late in the human analysis chain.

The acoustic transfer function of the human external ear is uniquely related to human body geometry. It is composed of four parts that can be modeled mathematically, as shown in Fig. 1.135, or recreated by an artificial head system.[15,16,17]

Reflections and diffraction of sound at the upper body, the shoulder, the head, and the outer ear (pinna), as well as resonances caused by the concha and ear canal, are mainly responsible for the transfer characteristic. The cavum concha is the antechamber to the ear. The spectral shape of the external ear transfer function varies from person to person due to the uniqueness of people and the

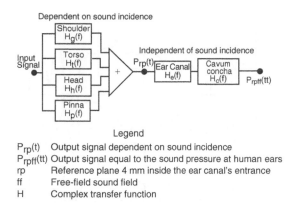

FIGURE 1.135   The human external-ear transfer function.

dimensions of these anatomical features. Therefore, both artificial heads and their mathematical models are based on statistical composites of responses and dimensions of a number of persons.

All of these contributions to the external ear transfer function are direction sensitive. This means that sound from each direction has its own individual frequency response. In addition, the separation of the ears with the head in between affects the relative arrival time of sounds at the ears. As a result the complete outer-ear transfer function is very complicated, Fig. 1.136, and can only be partially applied as a correction to the response of a single or even a pair of microphones. In the figure, the solid curve indicates reference SPL. The solid curves represent the free-field (direct sound) external ear transfer function, while the dashed curves represent the difference, at each direction, relative to frontal free-field sound incidence.

Artificial heads have been used for recording for some time. However, the latest heads and associated signal processing electronics have brought the state of the art close to in the ear (ITE) recording, which places microphones in human ears.

The KU 100 measurement and recording system by Georg Neumann GmbH in Germany is an example of a high-quality artificial head system, Fig. 1.137. Originally developed by Dr. Klaus Genuit and his associates at the Technical University of Aachen, the artificial head, together with carefully designed signal processing equipment, provides binaural recording systems that allow very accurate production of spatial imaging of complex sound fields.

The head is a realistic replica of a human head and depends on a philosophy of sound recording and reproduction—namely, that the sound to be recreated for a listener should not undergo two transfer functions, one in the ears of the artificial head and one in the ears of the listener.

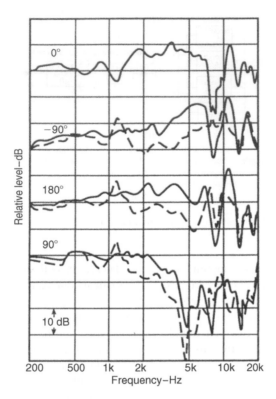

FIGURE 1.136 Transfer function of the left ear, measured 4 mm inside the entrance of the ear canal, for four angles of incidence (straight ahead, to the left, straight behind, and to the right).

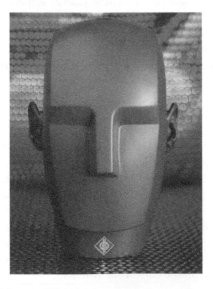

FIGURE 1.137 Georg Neumann KU 100 dummy head. Courtesy Georg Neumann GmbH.

Figure 1.138 is a block diagram of a head microphone and recording system. A high-quality microphone is mounted at the ear canal entrance position on each side of the head. Signals from each microphone pass through diffuse-field equalizers in the processor and are then available for further use in recording or reproduction. The diffuse-field equalizer is specifically tuned for the head to be the inverse of the frontal diffuse-field transfer function of the head. This signal is then recorded and can be used for loudspeaker playback and for measurement. The headphone diffuse-field equalizers in the Reproduce Unit yield a linear diffuse-field transfer function of the headphone, so the sound pressures presented at the entrance of the listener's ear canals will duplicate those at the entrance of the head's ear canals.

Diffuse-field equalization is suitable for situations where the source is at a distance from the head. For recordings close to a sound source or in a confined space, such as the passenger compartment of an automobile, another equalization called *independent of direction* (ID) is preferred. The equalization is internal in the head in Fig. 1.138.

Signals $P_{HR}(t)$ and $P_{HI}(t)$ from the heads can be recorded and used directly for loudspeaker playback, analysis, or playback through headphones. As a recording tool this method can surpasse many other recording techniques intended for loudspeaker reproduction. The full benefits of spatial imaging can be heard and enjoyed with earphone playback as well as with high quality loudspeakers.

The heads are constructed of rugged fiberglass. The microphones can be calibrated by removal of the detachable outer ears and applying a pistonphone. Preamplifiers on the microphones provide polarization and have balanced transformer less line drivers. A record processor and modular unit construction provides dc power to the dummy head and acts as the interface between the head and the recording medium or analysis equipment. The combination of low noise electronics and good overload range permits full use of the 135 dB dynamic range of the head microphones and 145 dB with the 10 dB attenuator switched in.

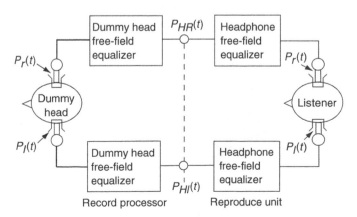

FIGURE 1.138   Block diagram of a dummy head binaural stereo microphone system.

For headphone playback, a reproduce unit provides an equalized signal for the headphones that produces ear canal entrance sound signals that correspond to those at the corresponding location on the artificial head.

An important parameter to consider in any head microphone recording system is the dynamic range available at this head signal output. For example, the canal resonance can produce a sound pressure that may exceed the maximum allowed on some ear canal mounted microphones.

1.8.2 In the Ear Recording Microphones[18]

In-the-Ear (ITE™) recording and Pinna Acoustic Response (PAR™) playback represent a new-old approach to the recording of two channels of spatial images with full fidelity and their playback over two channels, Fig. 1.138. It is important that the loudspeakers are in signal synchronization and that they be placed at an angle so that the listener position is free of early reflections.

Low noise, wide frequency, and dynamic range probe microphones employing soft silicone probes are placed in the pressure zone of the eardrum of live listeners. This microphone system allows recording with or without equalization to compensate for the ear canal resonances while leaving the high-frequency comb filter spatial clues unaltered. The playback system consists of synchronized loudspeaker systems spaced approximately equal distances from the listener in

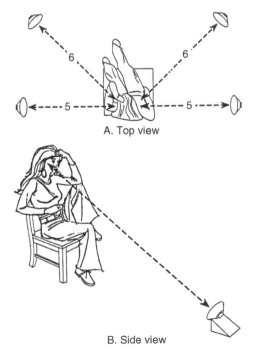

A. Top view

B. Side view

FIGURE 1.139   The loudspeaker arrangement used for PAR playback of ITE recordings. Courtesy Syn-Aud-Con.

the pattern shown in Fig. 1.139. Both left loudspeakers are in parallel, and both right loudspeakers are in parallel. However, the front and back loudspeakers are on individual volume controls. This is to allow balancing side-to-side and to adjust the front-to-back relative levels for each individual listener. The two front loud-speakers are used to provide hearing signals forward of the listener.

Figure 1.140A shows an ETC made in a listening room ($L_D - L_R = 0.24$). Figure 1.140B is the identical position measured with the ITE technique ($L_D - L_R = 5.54$). Note particularly the difference in $L_D - L_R$ for the two techniques.

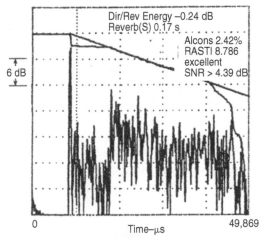

A. An ETC made in a listening room with a GenRAd 1/2 inch microphone ($L_D - L_R = 0.24$).

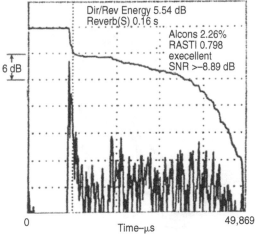

B. The identical position measured with the ITE technique ($L_D - L_R = 0.24$).

FIGURE 1.140   An ETC comparison of a measurement microphone and the ITE technique at the same position in the room. Courtesy Syn-Aud-Con.

ITE recording and PAR playback allow a given listener to hear a given speech intelligibility environment as perceived by another person's head and outer ear configuration right up to the eardrum.

Recordings made using ITE microphones in two different people's ears of the same performance in the same seating area sound different. Playback over loudspeakers where the system is properly balanced for perfect geometry for one person may require as much as 10 dB different front-to-back balance for another person to hear perfect geometry during playback.

Since ITE recordings are totally compatible with normal stereophonic reproduction systems and can provide superior fidelity in many cases, the practical use of ITE microphony would appear to be unlimited.

1.9 USB MICROPHONES

The computer has become an important part of sound systems. Many consoles are digital and microphones are connected directly to them. Microphones are also connected to computers through the USB input.

The audio-technica AT2020 USB cardioid condenser microphone, Fig. 1.141, is designed for computer based recording. It includes a USB (Universal Serial Bus) digital output that is Windows and Mac compatible. The sample rate is 44.1 kHz with a bit depth of 16 bits. The microphone is powered directly from the 5 Vdc USB output.

FIGURE 1.141   Audio-technica AT2020 USB microphone. Courtesy Audio-Technica U.S., Inc.

FIGURE 1.142   MXL.006 USB microphone. Courtesy Marshall Electronics, Inc.

The MXL.006 USB is a cardioid condenser microphone with a USB output that connects directly to a computer without the need for external mic preamps through USB 1.1 and 2.0, Fig. 1.142.

The analog section of the MXL.006 microphone features a 20 Hz–20 kHz frequency response, a large gold diaphragm, pressure-gradient condenser capsule, and a three-position, switchable attenuation pad with settings for Hi (0 dB), Medium (–5 dB), and Lo (–10 dB). The digital section features a 16 bit Delta Sigma A/D converter with a sampling rate of 44.1 kHz and 48 kHz. Protecting the instrument's capsule is a heavy-duty wire mesh grill with an integrated pop filter.

The MXL.006 includes a red LED behind the protective grill to inform the user that the microphone is active and correctly oriented. The MXL.006 ships with a travel case, a desktop microphone stand, a 10 ft USB cable, windscreen, an applications guide, and free downloadable recording software for PCs and Mac.

1.10 WIRELESS COMMUNICATION SYSTEMS

Wireless communication systems are wireless microphones (*radio microphones*), Fig. 1.143, and a related concept, wireless intercoms. Often the same end user buys both the microphones and intercoms for use in television and radio broadcast production, film production, and related entertainment-oriented applications.

Wireless microphone systems can be used with any of the preceding microphones discussed. Some wireless microphone systems include a fixed microphone cartridge while others allow the use of cartridges by various manufacturers.

A block diagram of a wireless microphone system is shown in Fig. 1.144. The sending end of a wireless microphone system has a dynamic, condenser, electret, or pressure zone microphone connected to a preamplifier, compressor, and a small transmitter/modulator and antenna.

The receiving end of the system is an antenna, receiver/discriminator, expander, and preamplifier, which is connected to external audio equipment.

In a standard intercom system, each person has a headset and belt pack (or equivalent), all interconnected by wires. Wireless intercoms are essentially identical in operation, only they use no cable between operators. Instead, each belt pack includes a radio transmitter and receiver. The wireless intercom user typically wears a headset (a boom microphone with one or two earpieces) and can simultaneously transmit on one frequency and receive on another. The wireless intercom transmitter is virtually identical to a wireless microphone transmitter, but the receiver is miniaturized so that it, too, can be conveniently carried around and operated with minimum battery drain.

Wireless microphones are widely used in television production. Handheld models (integral microphone capsule and transmitter) are used by performers "on camera," where they not only free the performer to walk around and gesture spontaneously, they also avoid the need for stage personnel to feed wires around

FIGURE 1.143   Shure UHF-R wireless microphone system. Courtesy Shure Incorporated.

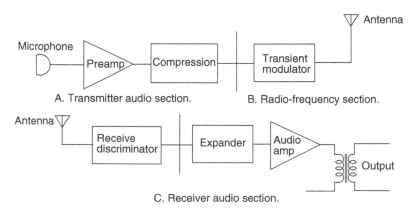

FIGURE 1.144   Wireless microphone transmitter section with built-in preamplifier, compressor, and transmitter, and the receiver with a built-in discriminator expander.

cameras, props, etc. Lavalier models (small pocket-sized transmitters that work with lavalier or miniature "hidden" microphones) are used in game shows, soap operas, dance routines, etc., where they eliminate the need for boom microphones and further avoid visual clutter.

For location film production, as well as electronic news gathering (ENG) and electronic field production (EFP), wireless microphones make it possible to obtain usable "first take" sound tracks in situations where previously, postproduction dialogue looping was necessary.

In theatrical productions, wireless microphones free actors to speak and/ or sing at normal levels through a properly designed sound-reinforcement system.

In concerts, handheld wireless microphones permit vocalists to move around without restriction, and without shock hazard even in the rain. Some lavalier models have high-impedance line inputs that accept electric guitar cords to create wireless guitars.

In all of the above applications where wireless microphones are used, in the studio or on location, a wireless intercom also is an invaluable communication aid between directors, stage managers, camera, lighting and sound crews, and security personnel. For cueing of talent and crews (or monitoring intercom conversations), economical receive-only units can be used. In sports production, wireless intercoms are used by coaches, spotters, players, production crews, and reporters. A major advantage is zero setup time. In critical stunt coordination, a wireless intercom can make the difference between a safe event or none at all.

1.10.1 Criteria for Selecting a Wireless Microphone

There are a number of criteria that must be considered in obtaining a wireless microphone system suitable for professional use.[19,20] Ideally, such a system must work perfectly and reliably in a variety of tough environments with good intelligibility and must be usable near strong RF fields, lighting dimmers, and sources of electromagnetic interference. This relates directly to the type of modulation (standard frequency modulation or narrow-band frequency modulation), the operating frequency, high-frequency (HF), very high-frequency (VHF), ultrahigh-frequency (UHF), the receiver selectivity, and so forth. The system must be very reliable and capable of operating at least five hours on one set of disposable batteries (or on one recharge if Ni-Cads are used).

1.10.1.1 Frequency Bands of Operation

Based on the FCC's reallocation of frequencies and the uncertainty of current and future allocations, some wireless manufacturers are offering systems that avoid the VHF and UHF bands completely. The ISM (industrial, science, and medicine) bands provide a unique alternative to the TV bands. By international agreement, all devices are low powered so there will never be any grossly high-powered RF interference potential. The 2.4 GHz band provides a viable alternative to traditional UHF bands, and as long as line of sight between transmitters and receivers is monitored users can easily get a 100 meter range. Another benefit of 2.4 GHz is that it can simplify wireless inventory for traveling shows. The same wireless frequencies are accepted worldwide, so there is no need to adhere to the country-specific frequency rules that severely complicate the situation for international tours. The same applies within the United States—the same frequencies work in all areas.

Currently wireless microphones are licensed on several frequencies, the most common being:

VHF low band (AM and FM)	25 to 50 MHz
	72 to 76 MHz
FM broadcast (FM)	88 to 108 MHz
VHF high band (FM)	150 to 216 MHz
UHF (FM)	470 to 746 MHz
	902 to 952 MHz

The VHF bands are seldom used anymore and can only be found on old equipment. The low band is in the noisiest radio spectrum and, because the wavelength is about 20 ft (6 m) it requires a long antenna (5 ft or 1.5 m). The VHF low band is susceptible to skip, which can be defined as external signals from a long distance away bouncing off the ionosphere back to earth, creating interference.

The VHF high band is more favorable than the low band. The ¼-wavelength antenna is only about 17 inches (43 cm) long and requires little space. The VHF band has some penetration through buildings that can be advantageous and disadvantageous. It is advantageous in being able to communicate between rooms and around surfaces. It is disadvantageous in that transmission is not controlled (security), and outside noise sources can reach the receiver.

Most often the frequencies between 174 MHz and 216 MHz are used in the VHF band, corresponding to television channels 7 to 13. The VHF high band is free of citizens band and business radio interference, and any commercial broadcast stations that might cause interference are scheduled so you know where they are and can avoid them. Inherent immunity to noise is built in because it uses FM modulation. Better VHF high-band receivers have adequate selectivity to reject nearby commercial television or FM broadcast signals. If operating the microphone or intercom on an unused television channel—for instance Channel 7—protection might be required against a local television station on Channel 8. Another problem could be caused by an FM radio station. If a multi–thousand watt FM station is broadcasting near a 50 mW wireless microphone, even a well-suppressed second harmonic can have an RF field strength comparable to the microphone or intercom signal because the second harmonic of FM 88 MHz is 176 MHz, which is in the middle of television Channel 7. The second harmonic of FM 107 MHz is 214 MHz, which is in the middle of

Channel 13. Thus, if a VHF wireless system is to be utilized fully, especially with several microphones or intercoms on adjacent frequencies, the wireless receiver must have a very selective front end.

One television channel occupies a 6 MHz wide segment of the VHF band. Channel 7, for example, covers from 174–180 MHz. A wireless intercom occupies about 0.2 MHz (200 kHz). By FCC Part 74 allocation, up to 24 discrete VHF high-band microphones and/or intercoms can be operated in the space of a single television channel. In order to use multiple systems on adjacent frequencies, the wireless microphone/intercom receivers must be very selective and have an excellent capture ratio (see Section 1.10.1.3). On a practical basis, this means using narrow-deviation FM (approximately 12 kHz modulation). Wide-deviation systems (75 kHz modulation or more) can easily cause interference on adjacent microphone/intercom frequencies; such systems also require wide bandwidth receivers that are more apt to be plagued by interference from adjacent frequencies. The trade-off between narrowband FM and wideband FM favor wideband for better overall frequency response, lower distortion, and inherently better SNR versus maximum possible channels within an unused TV channel for equal freedom from interference (max. 6). Poorly designed FM receivers are also subject to *desensing*. Desensing means the wireless microphone/intercom receiver is muted because another microphone, intercom, television station, or FM station (second harmonic) is transmitting in close proximity; this limits the effective range of the microphone or intercom.

The UHF band equipment is the band of choice and is the only one used by manufacturers today. The wavelength is less than 3 ft (1 m) so the antennas are only 9 inches (23 cm). The range is not as good as VHF, because it can sneak through small openings and can reflect off surfaces more readily.

All of the professional systems now are in the following UHF bands:

- A band 710–722 MHz.
- B band of 722–734 MHz
- 728.125–740.500 MHz band.

The FCC has assigned most of the DTV channels between channel 2 and 51, and only four channels between 64 and 69, which is where most of the professional wireless microphones operate.

1.10.1.2. Adjustment of the System's Operating Frequency

Many of the professional wireless microphones are capable of being tuned to many frequencies. In the past the systems were fixed frequency, often because that was the only way they could be made stable. With PLL-synthesized channels (Phase Lock Loop), it is not uncommon for systems to be switch tunable to 100 different frequencies in the UHF band and have a frequency stability of 0.005%. This is especially important with DTV coming into the scene.

1.10.1.3 Capture Ratio and Muting

Capture ratio and muting specifications of the receiver are important. The capture ratio is the ability of the receiver to discriminate between two transmitters transmitting on the same frequency. When the signal is frequency modulated (FM), the stronger signal controls what the receiver receives. The capture ratio is the difference in the signal strength between the capturing transmitter and the captured transmitter that is blanketed. The lower the number, the better the receiver is at capturing the signal. For instance, a receiver with a capture ratio of 2 dB will capture a signal that is only 2 dB stronger than the second signal.

Most systems have a muting circuit that squelches the system if no RF signal is present. To open the circuit, the transmitter sends a special signal on its carrier that breaks the squelch and passes the audio signal.

1.10.1.4 RF Power Output and Receiver Sensitivity

The maximum legal RF power output of a VHF high-band microphone or intercom transmitter is 50 mW; most deliver from 25–50 mW. Up to 120 mW is permissible in the business band (for wireless intercoms) under FCC part 90.217, but even this represents less than 4 dB more than 50 mW. The FCC does not permit the use of high-gain transmitter antennas, and even if they did, such antennas are large and directional so they would not be practical for someone who is moving around. Incidentally, high-gain receiving antennas are also a bad idea because the transmitter is constantly moving around with the performer so much of the received radio signal is actually caught on the bounce from walls, props, and so on (see Section 1.9.2).

Even if an offstage antenna is aimed at the performer, it probably would be aiming at the wrong target. Diversity receiving antenna systems, where two or more antennas pickup and combine signals to feed the receiver, will reduce dropouts or fades for fixed receiver installations.

The received signal level can't be boosted, given the restrictions on antenna and transmitted power, so usable range relies heavily on receiver sensitivity and selectivity (i.e., capture ratio and SNR) as well as on the audio dynamic range. In the pre-1980 time frame, most wireless microphones and intercoms used a simple compressor to avoid transmitter overmodulation. Today, systems include compandor circuitry for 15–30 dB better audio SNR without changing the RF SNR (see Section 1.10.3). This is achieved by building a full-range compressor into the microphone or intercom transmitter, and then providing complementary expansion of the audio signal at the receiver—much like the encoder of a tape noise-reduction system. The compression keeps loud sounds from overmodulating the transmitter and keeps quiet sounds above the hiss and static. The expander restores the loud sounds after reception and further reduces any low-level hiss or static. Companding the audio signal can provide from 80–85 dB of dynamic range compared to the 50–60 dB of a straight noncompanded transmit/receive system using the same deviation.

1.10.1.5 Frequency Response, Dynamic Range, and Distortion

No wireless microphone will provide flat response from 20 Hz–20 kHz, nor is it really needed. Wireless or not, by the time the audience hears the broadcast, film, or concert, the frequency response has probably been reduced to a bandwidth from 40 Hz–15 kHz. Probably the best criteria for judging a hand-held wireless microphone system is to compare it to the microphone capsule's naked response. If the transmit/receive bandwidth basically includes the capsule's bandwidth, it is enough. Generally speaking, a good wireless microphone should sound the same as a hard-wired microphone that uses the same capsule. Wireless intercom systems, because they are primarily for speech communication, are less critical with regard to audio bandwidth; 300 Hz–3 kHz is telephone quality, and 50 Hz–8 kHz is excellent for an intercom.

Dynamic range is probably the most critical aspect of performance for natural sound. A good compandor system will provide 80–85 dB of dynamic range, assuming the microphone is adjusted to 100% modulation on the loudest sounds. Leaving a margin of safety by turning down the microphone modulation level sacrifices SNR. Even with extra headroom and a working SNR of 75 dB, the microphone will still have about twice the dynamic range of a typical optical film sound track or television show.

The system should provide at least 40–50 dB SNR with a 10 μV signal and 70–80 dB SNR with an 80 μV signal. This shows up when no audio signal is being transmitted.

When an electret condenser microphone is used, a major limitation in dynamic range can be the capsule itself, not the wireless system. Typically, an electret powered by a 1.5 V battery is limited to about 105 dB SPL. Powered by a 9 V battery, the same microphone may be usable to 120 dB SPL. The wireless microphone system should be able to provide a high enough bias voltage to ensure adequate dynamic range from the microphone capsule. Although the condenser may be hotter in output level than a dynamic microphone, its background noise level is disproportionately higher, so the overall SNR specification may be lower.

Wireless intercom systems do not need the same dynamic range as a microphone. They do not have to convey a natural musical performance. However, natural dynamics are less fatiguing than highly compressed audio, especially given a long work shift. So aside from greater range, there are other benefits to seeking good SNR in the intercom: 40 dB or 50 dB would be usable, and 60 dB or 70 dB is excellent. An exception is in a very high-noise industrial environment, where a compressed loud intercom is necessary to overcome background noise. A good intercom headset should double as a hearing protector and exclude much of the ambient noise.

Distortion is higher in a wireless system than in a hard-wired system—a radio link will never be as clean as a straight piece of wire. Still, total harmonic distortion (THD) specifications of less than 1% overall distortion are available in today's better wireless microphones. In these microphones, one of the largest contributors to harmonic distortion is the compandor, so distortion is traded for SNR. The wireless intercom can tolerate more THD, but lower distortion will prevent fatigue and improve communication.

1.10.2 Receiving Antenna Systems

RF signal dropout or multipath cancellation is caused by the RF signal reflecting off a surface and reaching a single receiver antenna 180° out-of-phase with the direct signal, Fig. 1.145. The signal can be reflected off surfaces such as armored concrete walls, metal grids, vehicles, buildings, trees, and even people.

Although you can often eliminate the problem by experimenting with receiver antenna location, a more foolproof approach is to use a space diversity system where two or more antennas pickup the transmitted signal, as shown in Fig. 1.146. It is highly unlikely that the obstruction or multipath interference will affect two or more receiver antennas simultaneously.

There are three diversity schemes: switching diversity, true diversity, and antenna combination.

- **Switching Diversity.** In the switching diversity system, the RF signals from two antennas are compared, and only the stronger one is selected and fed to one receiver.

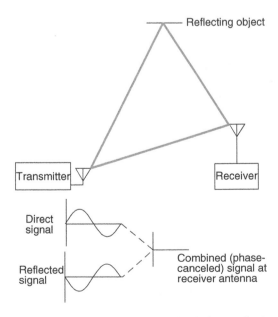

FIGURE 1.145   Phase cancellation of radio-frequency signals due to reflections.

- **True Diversity.** This receiving technique uses two receivers and two antennas set up at different positions, Fig. 1.147. Both receivers operate on the same frequency. The AF signal is taken from the output of the receiver that at any given moment has the stronger signal at its antenna. The probability of no signal at both antennas at the same time is extremely small. The advantages of diversity compared to conventional RF transmission are shown in

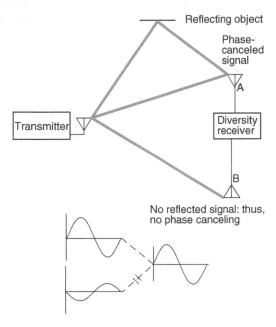

FIGURE 1.146 Diversity antenna system used to reduce multipath radio-frequency phase cancellation.

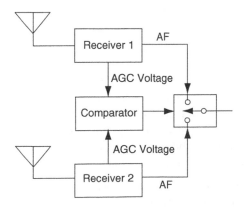

FIGURE 1.147 Functional diagram of a true diversity receiver.

Fig. 1.148. Only the receiving chain with the better input signal delivers audio output. Not only does this system provide redundancy of the receiving end, but it also combines signal strength, polarity, and space diversity.

- **Antenna Combination Diversity.** The antenna combination diversity system is a compromise of the other methods. This system uses two or more antennas, each connected to a wideband RF amplifier to boost the received signal. The signals from both receiving antennas are then actively combined and fed to one standard receiver per microphone. In this way, the receiver always gets the benefit of the signals present at all antennas. There is no switching noise, no change in background noise, and only requires one receiver for each channel. A drawback is the possibility of complete signal cancellation when phase and amplitude relationships due to multipath provide the proper unfavorable conditions.

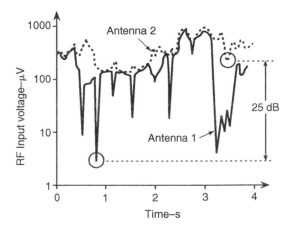

FIGURE 1.148   Effect of switch-over diversity operation. Solid line RF level at antenna 1 and the dotted line RF level at antenna 2. Courtesy Sennheiser Electronic Corporation.

1.10.2.1 Antenna Placement

It is often common to use a near antenna and a far antenna. The near antenna, which is the one nearest the transmitter, produces the majority of the signal most of the time; in fact, it may even be amplified with an in-line amplifier. The far-field antenna may be one or more antennas usually offset in elevation and position; therefore, the possibility of dropout is greatly reduced. Because the antennas are common to all receivers, many wireless microphones can be used at the same time on the same antenna system. This means that there are fewer antennas and a greater possibility of proper antenna placement.

The following will generally prevent dead spots:

- Do not set up antennas in niches or doorways.
- Keep the antennas away from metal objects including armored concrete walls. Minimum distance: 3 ft (1 m).

- Position the antennas as close as possible to the point where the action takes place.
- Keep antenna cables short to keep RF losses at a minimum. It is better to use longer AF leads instead.
  Note: If long runs of antenna cable are used, be sure they are of the low-loss type.
- Make a walk around test, i.e., operate the transmitter at all positions where it will be used later. Mark all points where field strength is weak. Try to improve reception from these points by changing the antenna position. Repeat this procedure until the optimum result is achieved.

Interference is mainly caused by spurious signals arriving at the receiver input on the working frequency. These spurious signals may have various causes:

- Two transmitters operating on the same frequency (not permissible).
- Intermodulation products of a multichannel system whose frequencies have not been selected carefully enough.
- Excessive spurious radiation from other radio installations—e.g., taxi, police, CB-radio, etc.
- Insufficient interference suppression on electric machinery, vehicle ignition noise, etc.
- Spurious radiation from electronic equipment—e.g., light control equipment, digital displays, synthesizers, digital delays, computers, etc.

1.10.3 Companding

Two of the biggest problems with using wireless microphones are SNR and dynamic range. To overcome these problems, the signal is compressed at the transmitter and expanded at the receiver. Figures 1.144 and 1.149 graphically illustrate how and what this can accomplish with respect to improving the SNR and reducing the susceptibility to low-level incidental FM modulation, such as buzz zones.

As the typical input level changes by a factor of 80 dB, the audio output to the modulator undergoes a contoured compression, so a change in input audio level is translated into a pseudologarithmic output. This increases the average modulation level, which reduces all forms of interference encountered in the transmission medium.

By employing standard narrowband techniques at the receiver, the recovered audio is virtually free of adjacent channel and spurious response interference. In addition, up to ten times the number of systems can be operated simultaneously without cross-channel interference. The ability of the receiver to reject all forms of interference is imperative when utilizing expansion and compression techniques because the receiver must complementarily expand the audio component to restore the original signal integrity.

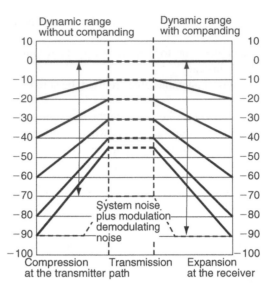

FIGURE 1.149 Compression and expansion of the audio signal. Notice the −80 dB signal is not altered, and the −20 dB signal is altered significantly.

1.10.4 Waterproof Wireless Microphone Systems

Wireless microphones that are worn are very useful for coaching all forms of athletics including swimming and aquatic aerobics. If the instructor always stays on the pool deck, a weatherproof system might be adequate. If the instructor is in the water, a completely submersible and waterproof system will be required.

Hydrophonics assembles a completely waterproof and submersible wireless microphone system. Assembled with Telex components, the system includes a headset microphone with a special waterproof connector and a Telex VB12 waterproof beltpack transmitter. The transmitter can operate on a 9 V alkaline battery or a 9 V NiMH rechargeable battery. The rechargeable battery is recommended as it does not require removing the battery from the transmitter for recharging and therefore reduces the chance of water leaking into the transmitter housing. The receiver is a Telex VR12 for out-of-pool operation, and can be connected to any sound system the same way as any other wireless microphone.

An interesting thing about this system is you can dive into the water while wearing the system and come up and immediately talk as the water drains out of the windscreen rapidly.

The DPA Type 8011 hydrophone, Fig. 1.150, is a 48 V phantom powered waterproof microphone specially designed to handle the high sound pressure levels and the high static ambient pressure in water and other fluids. The hydrophone uses a piezoelectric sensing element, which is frequency compensated to match the special acoustic conditions under water. A 10 m high-quality audio

FIGURE 1.150   DPA 8011 hydrophone. Courtesy DPA Microphones A/S.

cable is vulcanized to the body of the hydrophone and fitted with a standard three-pin XLR connector. The output is electronically balanced and offers more than 100 dB dynamic range. The 8011 hydrophone is a good choice for professional sound recordings in water or under other extreme conditions where conventional microphones would be adversely affected.

1.11 MULTICHANNEL WIRELESS MICROPHONE AND MONITORING SYSTEMS

By Joe Ciaudelli and Volker Schmitt

1.11.1 Introduction

The use of wireless microphones and monitoring systems has proliferated in the past few years. This is due to advancements in technology, a trend towards greater mobility on stage, and the desire to control volume and equalization of individual performers. Consequently, installations in which a number of wireless microphones, referred to as channels, are being used simultaneously, have increased dramatically. Now theatres and studios with large multichannel systems, greater than thirty channels, are common. Systems of this magnitude are a difficult engineering challenge. Careful planning, installation, operation, and maintenance are required.

Wireless systems require a transmitter, transmit antenna, and receiver to process sound via radio-frequency (RF) transmission. First, the transmitter processes the signal and superimposes it on a carrier through a process called *modulation*. The transmit antenna then acts as a launch pad for the modulated carrier and broadcasts the signal over the transfer medium: air. The signal must then travel a certain space or distance to reach the pickup element, which is the receiving antenna.

Finishing up the process, the receiver—which selects the desired carrier—strips off the signal through demodulation, processes it, and finally reconstitutes the original signal. Each wireless channel needs to operate on a unique frequency.

1.11.2 Frequencies

Manufacturers generally produce wireless microphones on ultrahigh frequencies (UHF) within the TV band with specifications outlined by government agencies such as the Federal Communications Commission (FCC). The wavelength is inversely proportional to the frequency. Higher frequencies have shorter wavelengths. UHF frequencies (450–960 MHz) have a wavelength of less than 1 meter. They have excellent reflective characteristics. They can travel through a long corridor, bouncing off the walls, losing very little energy. They also require less power to transmit the same distance compared to much higher frequencies, such as microwaves. These excellent wave propagation characteristics and low power requirements make UHF ideal for performance applications.

1.11.3 Spacing

In order to have a defined channel, without crosstalk, a minimum spacing of 300 KHz between carrier frequencies should be employed. A wider spacing is even more preferable since many receivers often exhibit desensitized input stages in the presence of closely spaced signals. However, caution should be used when linking receivers with widely spaced frequencies to a common set of antennas. The frequencies need to be within the bandwidth of the antennas.

1.11.4 Frequency Deviation

The modulation of the carrier frequency in an FM system greatly influences its audio quality. The greater the deviation, the better the high-frequency response and the dynamic range. The trade-off is that fewer channels can be used within a frequency range. However, since audio quality is usually the priority, wide deviation is most desirable.

1.11.5 Frequency Coordination

Multichannel wireless microphone systems can be especially difficult to operate, as they present several special conditions. Multiple transmitters moving around a stage will result in wide variations of field strength seen at the receiver antenna system. This makes frequency selection to avoid interference from intermodulation (IM) products highly critical. This is even more challenging in a touring application since the RF conditions vary from venue to venue. In this case, the mix of frequencies is constantly changing. The daunting task to coax each of these variables to execute clear audio transmission can be achieved through careful frequency coordination.

Intermodulation is the result of two or more signals mixing together, producing harmonic distortion. It is a common misconception that intermodulation is produced by the carrier frequencies mixing within the air. Intermodulation occurs within active components, such as transistors, exposed to strong RF input signals. When two or more signals exceed a certain threshold, they drive the active component into a nonlinear operating mode and IM products are generated. This usually happens in the RF section of the receiver, in antenna amplifiers, or the output amplifier of a transmitter. In multichannel operation, when several RF input signals exceed a certain level the IM products grow very quickly. There are different levels of intermodulations defined by the number of addition terms.

In any wireless system with three or more frequencies operating in the same range, frequency coordination is strongly advised.

It is necessary to consider possible IM frequencies that might cause problems for the audio transmission. The 3rd and 5th harmonics, in particular, might raise interference issues.

The following signals may be present at the output of a nonlinear stage:

Fundamentals:	F1 and F2
Second Order:	2F1, 2F2, F1±F2, F2−F1
Third Order:	3F1, 3F2, 2F1±F2, 2F2±F1
Fourth Order:	4F1, 4F2, 2F1±2F2, 2F2±2F1
Fifth Order:	5F1, 5F2, 3F1±2F2, 3F2±2F1
Additional higher orders	

As a result, the intermodulation frequencies should not be used, as those frequencies are virtual transmitters. The fundamental rule *never use two transmitters on the same frequency* is valid in this case. However, even-order products are far removed from the fundamental frequencies and, for simplicity, are therefore omitted from further considerations. Signal amplitude rapidly diminishes with higher-order IM products, and with contemporary equipment design, consideration of IM products can be limited to 3rd and 5th order only.

For multichannel applications such as those on Broadway (i.e., 30+ channels), the IM products can increase significantly and the calculation of intermodulation-free frequencies can be done by special software. By looking only at the third harmonic distortion in a multichannel system, the number of third-order IM products generated by multiple channels is:

- 2 channels result in 2.
- 3 channels result in 9.
- 4 channels result in 24.
- 5 channels result in 50.
- 6 channels result in 90.
- 7 channels result in 147.

- 8 channels result in 225.
- ....
- 32 channels result in 15,872 third-order IM products.

Adding more wireless links to the system will increase the number of possible combinations with interference potential logarithmically: n channels will result in $(n3 - n2)/2$ third-order IM products. Equal frequency spacing between RF carrier frequencies inevitably results in two- and three-signal IM products and must be avoided!

The RF level and the proximity define the level of the IM product. If two transmitters are close, the possibility of intermodulation will increase significantly. As soon as the distance between two transmitters is increased, the resulting IM product decreases significantly. By taking this into consideration, the physical distance between two or more transmitters is important. If a performer needs to wear two body pack transmitters, it is recommended to use two different frequency ranges and to wear one with the antenna pointing up and the other with it pointing down.

If the number of wireless channels increases, the required RF bandwidth increases significantly, Fig. 1.151.

External disturbing sources such as TV transmitters, taxi services, police services, digital equipment, etc., also have to be taken into consideration. Fortunately, the screening effect of buildings is rather high (30–40 dB). For indoor applications, this effect keeps strong outside signals at low levels. A significant problem can occur when poorly screened digital equipment is working in the same room. These wideband disturbing sources are able to interfere with wireless audio equipment. The only solution to this problem is to replace the poorly screened piece of equipment with a better one.

Other RF systems that have to be considered for compatibility are:

1. TV stations "On-Air."
2. Wireless intercoms.

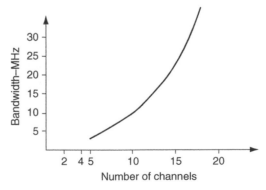

FIGURE 1.151 Bandwidth required for multichannel systems.

3. IFBs.
4. Wireless monitor systems.
5. Other wireless systems.

Compatibility between components of a system is achieved if the following requirements are met: each link in a multichannel wireless system functions equally well with all other links active and no single link—or any combination of multiple links—causes any interference.

If the transmitter of a wireless mic channel is switched off, its complementary receiver should also be switched off or muted at the mixing console. A receiver that does not see its transmitter will try to latch onto a nearby signal. That signal may be an IM product. The receiver will then try to demodulate this signal and apply it to the speaker system.

Equipment can be designed to minimize intermodulation. A specification known as *intermodulation rejection* or *suppression* is a measure of the RF input threshold before intermodulation occurs. For a well-designed receiver, this specification will be 60 dB or greater. An intermodulation rejection of 60 dB means that IM products are generated at input levels of approximately 1 mV. The highest quality multichannel receivers currently available feature an intermodulation rejection of >80 dB. If high-quality components are used, having an intermodulation suppression of 60 dB or greater, only the third-order products need to be considered.

1.11.6 Transmitter Considerations

Transmitters are widely available as portable devices, such as handheld microphones, body packs, and plug-on transmitters and are produced in stationary form as stereo monitors. When transmitting signals for most wireless applications via air, FM modulation is generally used; in doing so, one must also improve the sound quality in a variety of ways.

An RF transmitter works like a miniature FM radio station. First, the audio signal of a microphone is subjected to some processing. Then the processed signal modulates an oscillator, from which the carrier frequency is derived. The modulated carrier is radiated via the transmitter's antenna. This signal is picked up by a complementary receiver via its antenna system and is demodulated and processed back to the original audio signal.

1.11.6.1 Range and RF Power

Transmitter power is a rating of its potential RF signal strength. This specification is measured at the antenna output. The range of a wireless transmission depends on several factors. RF power, the operating frequency, the setup of the transmitter and receiver antennas, environmental conditions, and how the transmitter is held or worn are all aspects that determine the overall coverage of the system. Therefore, power specifications are of only limited use in assessing a transmitter's range, considering these variable conditions. Also, battery life is associated with RF output power. Increased power will reduce battery life with only a moderate increase in range.

Using RF wireless microphone transmitters with the right amount of RF output power is important to ensure total system reliability. There is a common misconception that higher power is better. However, in many applications high power can aggravate intermodulation (IM) distortion, resulting in audible noises.

First of all, the applied RF output power must fall within the limit allowed by each country's legislation. In the United States, the maximum RF output power for wireless microphones is limited to 250 mW. In most of the countries in Europe this figure is 50 mW, while in Japan it is only 10 mW. Despite the 10 mW limitation, many multichannel wireless microphones are operating in Japan. This is achieved by careful attention to factors like antenna position, use of low loss RF cables and RF gain structure of the antenna distribution system.

There are indeed some applications in which more RF output power is an appropriate measure; a perfect example would be a golf tournament, as the wireless system needs to cover a wide area. There are usually only a few wireless microphones in use at this type of function, and those microphones are generally not in close proximity to each other.

If transmitters with high RF power are close together, intermodulation usually occurs. At the same time, the RF noise floor in the performance area is increased. As a matter of fact, a transmitter in close proximity to another transmitter will not only transmit its own signal, but it will also receive the signal and add this to the RF amplifier stage.

1.11.6.2 Dc-to-Dc Converter

Transmitters should be designed to provide constant RF output power and frequency deviation throughout the event being staged. This can be achieved through the use of a dc-to-dc converter circuit. Such a circuit takes the decaying battery voltage as its input and regulates it to have a constant voltage output. Once the voltage of the batteries drops below a minimum level, the dc-to-dc converter shuts off, almost instantaneously. The result is a transmitter that is essentially either off or on. While it is on, the RF output power, frequency deviation, and other relevant specifications remain the same. Transmitters without regulation circuits, once the battery voltage begins to drop, will experience reduced range and audio quality.

1.11.6.3 Audio Processing

To improve the audio quality, several measures are necessary because of the inherent noise of the RF link.

1.11.6.3.1 Pre- and De-Emphasis

This method is a static measure that is used in most of the FM transmissions. By increasing the level of the higher audio frequencies on the transmitter side, the signal-to-noise ratio is improved because the desired signal is above the inherent noise floor of the RF link.

1.11.6.3.2 Companding

The compander is a synonym for *compressor* on the transmitter side and for *expander* on the receiving end. The compressor raises low audio level above the RF noise floor. The expander does the mirror opposite and restores the audio signal. This measure increases the signal-to-noise ratio to CD quality level.

1.11.6.3.3 Spurious Emissions

Apart from the wanted carrier frequency, transmitters can also radiate some unwanted frequencies known as spurious emissions. For large multichannel systems these spurious frequencies cannot be ignored. They can be significantly reduced through elaborate filtering and contained by using a well-constructed, RF tight metal housing for the transmitter. Also, an RF tight transmitter is less susceptible to outside interference.

A metal housing is important not only for its shielding properties, but also its durability. These devices usually experience much more abuse by actors and other talent than anyone ever predicts.

1.11.6.4 Transmitter Antenna

Every wireless transmitter is equipped with an antenna, which is critically important to the performance of the wireless system. If this transmitter antenna comes in contact with the human body, the transmitted wireless energy is reduced and may cause audible noises known as *dropouts*. This effect of detuning the antenna on contact is called body absorption.

For this reason, talent should not touch the antenna while using handheld microphones. Unfortunately, there is no guarantee that they will follow this recommendation. Taking this into account, optimized antenna setup at the receiver side and the overall RF gain structure of the system becomes critical.

This same effect can occur when using body pack transmitters, especially if the talent is sweating. A sweaty shirt can act as a good conductive material to the skin. If the transmitter antenna touches it, reduced power and thus poor signal quality may result. In this case, a possible approach is to wear the body pack upside down near or attached to the belt, with the antenna pointing down. Sometimes this measure does not work because the talent will sit on the antenna. In this case, a possible solution is keeping the transmitter in the normal position and fitting a thick-walled plastic tube over the antenna, such as the type that is used for aquarium filters.

1.11.7 Receiver Considerations

The receiver is a crucial component of wireless audio systems, as it is used to pick the desired signal and transfer its electrical information into an audio signal. Understanding basic receiver design, audio processing, squelch, and diversity operation can help ensure optimum performance of the system.

Virtually all modern receivers feature super heterodyne architecture, in which the desired carrier is filtered out from the multitude of signals picked up by the antenna, then amplified and mixed with a local oscillator frequency to generate the difference: *intermediate frequency* (IF). This IF undergoes more controlled discrimination and amplification before the signal is demodulated and processed to restore the output with all the characteristics and qualities of the original.

Audio signal processing of a receiver is the mirror opposite of the transmitter. Processing done in the transmitters often includes pre-emphasis (boosting high audio frequencies) as well as compression. These are reversed in the receiver by the de-emphasis and the expander circuit.

An inherent RF noise floor exists in the air. The squelch setting should be set above this noise level. This acts as a noise gate that mutes the audio output if the wanted RF signal falls below a threshold level. This prevents a blast of white noise through the PA if the RF signal is completely lost. If the squelch setting is too low, the receiver might pick the noise floor and this noise can be heard. If the squelch setting is too high the range of the wireless microphone is reduced.

1.11.7.1 RF Signal Level

Varying RF signal strength is mainly due to multipath propagation, absorption, and shadowing. These are familiar difficulties also experienced with car radios in cities.

Audible effects due to low RF signals, known as *dropouts*, can occur even at close range to the receiver due to multipath propagation. Some of the transmitted waves find a direct path to the receiver antenna and others are deflected off a wall or other object. The antenna detects the vector sum, magnitude, and phase of direct and deflected waves it receives at any particular instant. A deflected wave can diminish a direct wave if it has different phase, resulting in an overall low signal. This difference in phase is due to the longer path a deflected wave travels between the transmitter and receiver antennae and any phase reversal occurring when it hits an object. This phenomenon needs to be addressed in an indoor application since the field strength variation inside a building with reflecting walls is 40 dB or more. It is less critical outside.

RF energy can be absorbed by nonmetallic objects resulting in low signal strength. As stated previously, the human body absorbs RF energy quite well. It is important to place antennas correctly to minimize this effect.

Shadowing occurs when a wave is blocked by a large obstacle between the transmitter and receiver antennas. This effect can be minimized by keeping the antennas high and distance of ½ wavelength away from any large or metallic objects.

These problems are addressed by a diversity receiver. A diversity system is recommended even if only one channel is in operation. Large multichannel systems are only possible with diversity operation.

There are different kinds of diversity concepts available. Antenna switching diversity uses two antennas and a single receiving circuit. If the level at one antenna falls below a certain threshold it switches to the other antenna. This is an economical architecture but it leaves the chance that the second antenna could be experiencing an even lower signal than the one that falls below the threshold level. Another approach is the switching of the audio signal of two independent receiver units where each receiver unit is connected to its own antenna. This is known as *true diversity*. This technique improves the effective RF receiving level by greater than 20 dB. Depending on the diversity concept, active switching between the two antennas is a desired result.

The minimum distance between the two diversity antennas is very often an issue of debate. A minimum of ¼ of a wavelength of the frequency wave seems to be a good approach. Depending on the frequency, 5–6 inches is the minimum distance. In general, a greater distance is preferred.

1.11.8 Antennas

The position of the antenna and the correct use of its related components—such as the RF cable, antenna boosters, antenna attenuators, and antenna distribution systems—are the key to trouble-free wireless transmission. The antennas act as the eyes of the receiver, so the best results can be achieved by forming a direct line of sight between the transmitter antenna and receiver antenna of the system.

Receiving and transmitting antennas are available as omnidirectional and directional variants.

For receiving, omnidirectional antennas are often recommended for indoor use because the RF signal is reflected off of the walls and ceiling. When working outside, one should choose a directional antenna since there are usually little to no reflections outdoors, and this directivity will help to stabilize the signal. In general, it is wise to keep an antenna toolbox that contains both omnidirectional and directional antennas for use in critical RF situations, since they transmit and receive signals differently.

Omnidirectional antennas transmit or receive the signal by providing uniform radiation or response only in one reference plane, which is usually the horizontal one parallel to the earth's surface. The omnidirectional antenna has no preferred direction and cannot differentiate between a wanted and an unwanted signal.

If a directional antenna is used, it will transmit or receive the signal in the path it is pointing toward. The most common types are the yagi antenna and the log-periodic antenna, which are often wide-range frequency antennas covering the whole UHF range. In an outdoor venue, the desired signal can be received and the unwanted signal from a TV station can be rejected to a certain degree by choosing the correct antenna position. A directional antenna also transmits or receives only in one plane, like an omnidirectional antenna.

Several types of omnidirectional and directional antennas also exist for specific conditions. The telescopic antenna is an omnidirectional antenna and often achieves a wide range (450–960 MHz). If telescopic antennas are in use they should be placed within the line of sight of the counterpart antenna. They should not, for example, be mounted inside a metal flight case with closed doors as this will reduce the RF field strength from the transmitter and compromise the audio quality.

System performance will be raised considerably when remote antennas are used. A remote antenna is one that is separated from the receiver or transmitter unit. These antennas can be placed on a stand such as that for a microphone. This will improve the RF performance significantly. However, when using remote antennas, some basic rules need to be considered. Once again, a clear line of sight should be established between the transmitter and receiver antenna, Fig. 1.152.

Placing antennas above the talent increases the possibility the transmitter and receiver remain within line of sight, ensuring trouble-free transmission.

If a directional antenna is used, the position of the antenna and the distance to the stage is important. One common setup is pointing both receiving antennas toward the center of the stage. Once again, a line of sight between the receiver and transmitter antennas is best for optimum transmission quality.

Directional and omnidirectional antennas do have a preferred plane, which is either the horizontal or vertical plane. If the polarization between the transmitter and receiver antenna is different, this will cause some significant loss of the RF level. Unfortunately, it is not possible to have the same polarization of the antennas all of the time. In a theatrical application, the antenna is in a vertical position when the actress or actor walks on the stage. The polarization of the transmitter may change to the horizontal position if a scene requires the talent to lie down or crawl across the stage. In this case, circular polarized antennas can help. These kinds of antennas can receive the RF signal in all planes with the same efficiency.

Because the polarization of the antenna is critical and telescopic antennas are often used, it is not recommended to use the receiver antennas strictly in a horizontal or vertical plane. Rather, angle the antennas slightly as this will minimize the possibility that polarization would be completely opposite between transmitter and receiver.

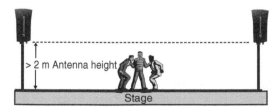

FIGURE 1.152 Placing antennas above the talent increases the possibility the transmitter and receiver remain within line of sight, ensuring trouble-free transmission.

One last note: The plural form for the type of antenna discussed in this article is *antennas*. Antennae are found on insects and aliens.

1.11.8.1 Antenna Cables and Related Systems

Antenna cables are often an underestimated factor in the design of a wireless system. The designer must choose the best cable for practical application, depending on the cable run and the installation, Table 1.3. As the RF travels down the cable its amplitude is attenuated. The amount of this loss is dependent on the quality of the cable, its length, and the RF frequency. The loss increases with longer cable and higher frequencies. Both of these effects must be considered for the design of a wireless microphone system.

RF cables with a better specification regarding RF loss are often thicker. These are highly recommended for fixed installations. In a touring application, in which the cable must be stored away each day, these heavier cables can be very cumbersome.

As any RF cable has some RF attenuation, cable length should be as short as possible without significantly increasing the distance between the transmitter and receiver antennas. This aspect is important for receiving applications but is even more critical for the transmission of a wireless monitor signal.

TABLE 1.3 Different Types of RF Cables with Various Diameters and the Related Attenuation for Different Frequencies

Cable Type	Frequency (MHz)	Attenuation (db/100')	Attenuation (dB/100 m)	Cable Diameter (inches/mm)
RG-174/U	400	19.0	62.3	0.110 / 2.8
	700	27.0	88.6	
RG-58/U	400	9.1	29.9	0.195 /
	700	12.8	42.0	4.95
RG-8X	400	6.6	21.7	0.242 / 6.1
	700	9.1	29.9	
RG-8/U	400	4.2	13.2	0.405 /
	700	5.9	19.4	10.3
RG-213	400	4.5	14.8	0.405 /
	700	6.5	21.8	10.3
Belden 9913	400	2.7	8.9	0.405 /
	700	3.6	11.8	10.3
Belden 9913F 9914	400	2.9	9.5	0.405 /
	700	3.9	12.8	10.3

Source: Belden Master Catalogue.

In a receiving application, it is important to consider losses from the cable as well as from any splitter in the antenna system during the design and concept stage of a wireless microphone system. If the losses in the system are small, an antenna booster should not be used. In this case, any dropout is not related to the RF loss in the antenna system; instead, it is more often related to the antenna position and how the transmitter is used and worn during the performance. An antenna booster is recommended if the loss in the antenna system is greater than 6 dB.

If an antenna booster is necessary, it should be placed as close as possible to the receiving antenna. Antennas with a built-in booster are known as *active antennas*. Some of these have a built-in filter, only allowing the wanted frequency range to be amplified. This is recommended because it reduces the possibility of intermodulation.

Two antenna boosters should not be used back-to-back when the RF cable run is very long. The second antenna booster would be overloaded by the output of the first amplifier and would produce intermodulation.

Special care must be taken when using an antenna booster if the transmitter comes close to the receiver antenna. The resulting strong signal could drive the antenna booster past its linear operation range, thus producing intermodulation products. It is recommended to design and install a system such that the transmitter remains at least 10 feet from the receiver antenna at all times.

Another important factor is the filter at the input stage of the antenna booster. The approach is to reduce the amount of unwanted signals in the RF domain as much as possible. This is another measure to reduce the possibility of intermodulation of this amplifier.

Also, signals that come from a TV station—such as a digital television (DTV) signal—are unwanted signals and can be the reason for intermodulation products in the first amplifier.

If the free TV channel between the DTV should be used for wireless microphone transmission, the DTV signals might cause the problems. To reduce the effect of DTV signals, a narrow input filter will help to overcome the possible effect of intermodulation.

Often a narrower filter at the input stage of a wireless receiver is preferable. This will often work for fixed installations because there are decreased possibilities that the RF environment will change. This is especially the case when the RF environment is difficult and a lot of TV stations or other wireless systems are operating.

1.11.8.2 Splitter Systems

Antenna splitters allow multiple receivers to operate from a single pair of antennas. Active splitters should be used for systems greater than four channels so that the amplifiers can compensate for the splitter loss. Security from interference and intermodulation can be enhanced by filtering before any amplifier stage. As an example, a 32-channel system could be divided into

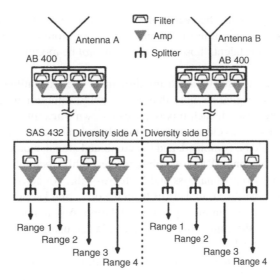

FIGURE 1.153   Diversity antenna setup with filtered boosters, long antenna cables, and active splitter with selective filtering.

four subgroups of eight channels. The subgroups can be separated from each other by highly selective filters. The subgroups can then be considered independent of each other. In this way, frequency coordination only needs to be performed within each group. It is much easier to coordinate eight frequencies four times than to attempt to coordinate a single set of 32 frequencies, Fig. 1.153.

1.11.9 Wireless Monitor Systems

Wireless monitor systems are essential for stage-bound musical productions. Perhaps the biggest advantage of a wireless monitor system is the ability to use an individual monitor mix for each musician on stage. Furthermore, a wireless monitor system significantly reduces the amount of, or even eliminates, monitor speakers in the performance area. This results in lower risk of feedback and a more lightweight, compact monitor system.

Some special precautions must be taken before using wireless monitor systems. In most cases, this signal is a stereo signal. This multiplexed signal is more sensitive to dropouts and static and multipath situations. For long range applications, mono operation can improve system performance.

If wireless microphones and wireless monitor systems are used in parallel, those systems should be separated in a way that the frequencies are at least 8 MHz apart and that the physical distance between the transmitter and the receiver is maximized. This will reduce the risk of blocking—an effect that desensitizes a receiver and prevents the reception of the desired signal. Therefore, if a body pack wireless mic transmitter and a wireless monitor receiver are both attached to the same talent, those devices should *not* be mounted directly beside each other.

When musicians use the same monitor mix, one transmitter can be used to provide the RF signal to more than one wireless monitor receiver. If individual mixes are desired, each mix requires its own transmitter operating on a unique frequency. To avoid intermodulation disturbances, the wireless monitor transmitters should be combined, and the combined signal should then be transmitted via one antenna. Active combiners are highly recommended. Passive combiners suffer from signal loss and high crosstalk. An active combiner isolates each transmitter by around 40 dB from the other and keeps the RF level the same (0 dB gain), thus minimizing intermodulation. Again, intermodulation is a major issue within the entire wireless concept. When using stereo transmission, it is even more critical.

When considering an external antenna, one important factor must be taken into consideration: the antenna cable should be as short as possible to avoid losses via the RF cable. A directional external antenna is recommended to reduce multipath situations from reflections, and it will have some additional passive gain that will increase the range of the system.

If remote antennas are used for the wireless monitor transmitters as well as wireless mic receivers, those antennas should be separated by at least 10–15 feet. Blocking of the receivers, as discussed above, is then avoided. Furthermore, the antennas should not come in direct contact with the metal of the lighting rig. This will detune the antenna and reduce the effective radiated wireless signal.

1.11.10 System Planning for Multichannel Wireless Systems

When putting together a multichannel wireless microphone system, several items are essential for trouble-free operation. First, you must understand the environment in which the system will be used.

Location. The location of a venue can be determined by using mapping tools on the Internet, such as Google Earth. If you figure out the coordinates of the venue, simply plug this information into the FCC homepage, http://www.fcc. gov/fcc-bin/audio/tvq.html. The result shows all transmitters licensed by the FCC in this area. This information will allow the designer of the wireless system to plan which vacant TV channels can be used for wireless audio devices. If there is a TV transmitter close to the location of the wireless microphone system (<70 miles), this TV channel should generally be avoided. Once one knows which TV channels may be used in the area, the designer can use another software tool that calculates the IM-free frequencies and displays possible setups.

Quantity and Frequency Coordination. Determine how many wireless microphones, wireless monitor systems, intercoms, etc. are required or in use for your job. With the information you gathered from step one, you can begin the system design. You now have the available TV channels and the number of wireless systems you want to use.

With this know-how you can start the frequency coordination of your system inside the vacant TV channels. This is supported by software that is available from various companies. The key here is to prevent intermodulation products (unwanted frequencies generated by harmonic distortion) from interfering with the wanted frequencies of your wireless systems.

A check in the venue is also necessary. If you have the chance, scout the location with a spectrum analyzer, Fig. 1.154. With this tool, you can verify that the information from the Internet is correct. Alternately, you can scroll through the tunable frequencies of your wireless receivers to scan the RF activity in the venue. Many receivers also have an auto scan function to find open frequencies. This cross-check is necessary to find out whether other wireless devices are in use that you do not have on your list, which could interfere with your signal during operation.

Tune Your Components. Set your individual transmitters and corresponding receivers to their coordinated frequencies. Switch on all components and perform a final test of compatibility. Physically space the transmitters a couple feet apart and at least 10 feet from the receiving antenna. Listen for any interference. Compatibility between components of a system is achieved if the following requirements are met: each link in a multichannel wireless system functions equally well with all other links active and no single link—or any combination of multiple links—causes interference.

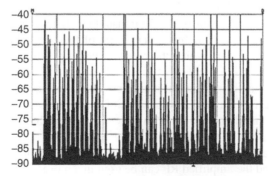

FIGURE 1.154 Plot of the RF spectrum in Athens outside the Olympic Stadium (450–960 MHz).

1.11.11 Future Considerations: Digital Wireless Transmission

Digital is a buzz word that many presume solves all the technical issues we face today. More and more digital equipment, such as mixing consoles, audio signal processors, and the like, are used for several applications, as a digital audio signal chain offers many advantages. A digital signal on a wire (i.e., fiber optic cable) is easier to handle than on a copper wire because 48, 64, or more audio channels can be transported on one thin fiber optic cable. If an audio signal is already in the digital domain, it makes sense to keep it in this domain as long as possible.

As for digital wireless transmission, a digital *wireless* system is beneficial when the sound, occupied RF spectrum, and battery lifetime is as good or even better than an analog system. On top of this, latency (time delay between input and output) is always a very important topic to keep in mind.

1.11.11.1 Starting with Sound and the Related Data Rate

The best sound can be expected if there is no audio data compression used in the wireless system. This will lead to a very high data rate.

- Minimum for 20 kHz audio and approximately 110 dB dynamic range: 18 bit $\times$ 48 kHz = 0.864 Mbit/s.
- Necessary overhead (framing, channel coding) leads to even higher data rate (factor approximately 1.5 [1.296 Mbit/s]).

When transmitting this high amount of data, it is no longer possible to use a simple and robust digital modulation scheme like FSK (frequency shift keying), ASK (amplitude shift keying), or PSK (phase shift keying), because these concepts will not be able to fulfill the spectrum mask, $\leq$ 200 kHz of occupied RF spectrum, defined by the FCC. Even if this constraint didn't exist, greater occupied RF spectrum could inhibit large multichannel systems.

To improve this, it is necessary to use a more complex modulation scheme with narrow filtering, Fig. 1.155.

The amplitude and the phase of the transmitted signal must be very precise when using this approach. Behind every point of the constellation diagram, a digital word is deposited, which the receiver has to pickup and transfer back into an audio signal.

This requires a very linear RF amplifier. This is a current-hungry device. The unwanted effect is reduced battery life of transmitters and portable receivers. By driving the RF amplifier with a better efficiency, the occupied RF spectrum will increase in an undesirable manner.

If the data rate described above can be reduced, the modulation scheme can be simplified and the amplified RF can be used in a more efficient way to conserve battery power and increase operational time.

To reduce the amount of digital data a compression algorithm has to be defined. This algorithm will add some latency to the whole data transmission

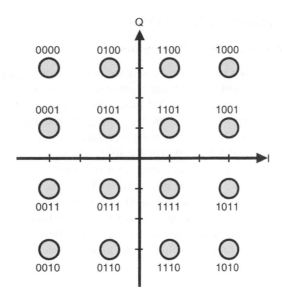

FIGURE 1.155   Constellation diagram of a 16 QAM modulation.

process. Low latency is especially important during a live performance on stage. If the total latency in a PA system, including contributions from digital mixing consoles, effects, etc., is >10 ms, the timing of the band will be thrown off. Furthermore, if streaming video is projected to accommodate a large audience the picture and sound will be out of sync.

New audio data compression algorithms show good performance with a very low latency. However, audio compression would introduce the possibility of audible artifacts (at least with awkward signals).

As technology improves, there will be solutions to the obstacles described above and digital will become available for wireless transmission.

The key questions for a digital system at this time are:

- Is data compression used?
- What RF spectrum is necessary and how will this impact multichannel systems?
- What is the latency of the system?
- What is the battery lifetime?

1.11.12 Conclusion

Large multichannel wireless systems demand excellent planning, especially in the initial phase, and good technical support. Observing all the above-mentioned items, perfect operation of a system can be guaranteed, even under difficult conditions.

1.12 MICROPHONE ACCESSORIES

1.12.1 Inline Microphone Processors

The overall sound of a microphone can often benefit from signal processing, and most mixers provide some basic equalization as a tool for customizing the sound of the microphone. Digital mixers provide an even greater set of tools, including parametric EQ, compression, gain management, and other automated functions. Dedicated signal processing for each microphone in a system provides a real advantage for the user, and some manufacturers are offering this sort of custom processing on a per-microphone basis via processors that plug inline with the microphone. These include automatic gain control, automatic feedback control, control for plosives and the proximity effect, and integrated infrared gates that turn the microphone on and off based on the presence of a person near the microphone. These phantom-powered processors allow for targeted solutions to many problems caused by poor microphone technique.

1.12.1.1 Sabine Phantom Mic Rider Pro Series 3

The series 3 Mic Rider includes infrared gates that turn the microphone on and off based on the presence of a person. The heat-sensing IR module is mounted on the gooseneck or is built in on the handheld version. The IR sensor can be adjusted for both time to turn off of 5–15 s and for distances of 3–9 ft, Figs. 1.156 and 1.157.

FIGURE 1.156   The Sabine Inline Mic Rider. Courtesy Sabine, Inc.

FIGURE 1.157 The Sabine gooseneck Mic Rider with built-in IR sensor. Courtesy Sabine, Inc.

1.12.1.2 Sabine Phantom Mic Rider Pro Series 2

The series 2 Mic Rider includes the adjustable IR Gate plus three audio processors: automatic gain control, proximity effect control for controlling increased bass due to the proximity effect, and plosive control for reducing pops and bursts from certain consonant sounds in speech.

1.12.1.3 Sabine Phantom Mic Rider Series 1

The series 1 Mic Rider includes Sabine's patented FBX Feedback Exterminator for maximizing gain before feedback plus the automatic gain control, proximity effect control, and plosive control. A nonadjustable IR gate is also included.

The Phantom Mic Rider works with 48 Vdc phantom power sources that conform to industry standards (DIN standard 45 596 or IEC standard 268–15A). Devices that do not conform can be modified to meet the standard, or external phantom power supplies can be used.

1.12.1.4 Lectrosonics UH400A Plug-On Transmitter

The design of this transmitter was introduced in 1988, in a VHF version aimed at broadcast ENG applications at a time when production crews were being downsized. Converting the popular dynamic microphones of the day to wireless operation eliminated the cable, which was very useful for the two-person production crews that had evolved. During the 20 years that followed, the design continued to evolve to address an ever-increasing variety of applications. The addition of selectable bias voltage allowed the transmitter to power electret microphones. The move to UHF frequencies and a dual-band compandor increased operating range and audio quality. Modifications to the design continued through the present day, leading to the current DSP-based model available in two versions for use with all types of microphones and modest line level signal sources.

The UH400A model has a 12 dB/octave low-frequency roll-off down 3 dB at 70 Hz. The UH400TM model offers an extended low frequency response down 3 dB at 35 Hz, Fig. 1.158. Figure 1.159 is the block diagram of the transmitter.

The most common applications of this transmitter are eliminating the cable between a microphone and a sound or recording system. A prime example

FIGURE 1.158   Lectrosonics UH400A plug-on transmitter. Courtesy Lectrosonics, Inc.

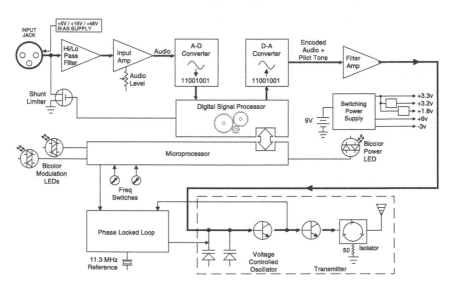

FIGURE 1.159   Block diagram of the Lectrosonics UH400A transmitter. Courtesy Lectrosonics, Inc.

is acoustic analysis in a large auditorium or stadium where measurements must be made at multiple locations around the sound system coverage area and extremely long cable runs are not practical. In this case, the wireless not only speeds up the process of making measurements, but it also allows more measurements to be taken, which can improve the final sound system performance.

Digital Hybrid Wireless™ is a patented process that combines a 24 bit digital audio stream with wide deviation FM (U.S. Patent 7,225,135). The process eliminates a compandor to increase audio quality and expand the applications to test and measurement and musical instrument applications.

Audio is sampled at 88.2 kHz and converted to a 24 bit digital stream. The DSP applies an encoding algorithm that creates what might be likened to an instruction set that is sent to the receiver via an FM carrier. The DSP in the receiver then applies an inverse of the encoding algorithm and regenerates the 24 bit digital audio stream.

An additional benefit of the FM radio link is the ability of the DSP to emulate a compandor for compatibility with analog receivers from Lectrosonics and two other manufacturers.

In the native hybrid mode, the FM deviation is ±75 kHz to provide a wide dynamic range. This wide deviation combined with 100 mW of output power provides a significant improvement in the audio SNR and the suppression of RF noise and interference.

Used with a microphone, the antenna is a dipole formed between the transmitter housing and the microphone body. When plugged into a console or mixer output, the housing of the transmitter is similar to the radiator of a ground plane antenna, with the console or mixer chassis functioning as the ground.

Phantom power can be set to 5, 15, or 48 V or turned off. The transmitter can provide up to 15 mA of current in 5 and 15 V settings, and up to 7 mA in the 48 V setting, allowing it to be used with any type of microphone, including high-end studio condenser models.

The transmitter is available on 9 different frequency blocks in the UHF band between 470 and 692 MHz. Each block provides 256 frequencies in 100 kHz steps.

1.12.1.5 MXL Mic Mate™ USB Adapter

The MXL Mic Mate™, Fig. 1.160, is a USB adapter used to connect a microphone to a Macintosh or PC computer. It uses a 16 bit Delta Sigma A/D converter with $THD + N = 0.01\%$ at sampling rates of 44.1 and 48.0 kHz and includes a three-position analog gain control. It includes a studio-quality USB microphone preamp with a balanced low noise analog input, supplies 48 Vdc phantom power to the microphone, and includes MXL USB Recorder Software for two track recording. There are three different Mic Mates, one for condenser microphones, one for dynamic microphones, and one for news line feeds, video cameras, etc.

FIGURE 1.160   MXL Mic Mate™ USB adapter. Courtesy Marshall Electronics.

1.12.2 Windscreens and Pop Filters

A *windscreen* is a device placed over the exterior of a microphone for the purpose of reducing the effects of breath noise and wind noise when recording out of doors or when panning or gunning a microphone. A windscreen's effectiveness increases with its surface area and the surface characteristics. By creating innumerable miniature turbulences and averaging them over a large area, the sum approaches zero disturbance. It follows that no gain is derived from placing a small foam screen inside a larger blimp type, Fig. 1.161A, whereas a furry cover can bring 20 dB improvement, Fig. 1.161B. Most microphones made today have an integral windscreen/pop filter built in. In very windy conditions, these may not be enough; therefore, an external windscreen must be used.

With a properly designed windscreen, a reduction of 20–30 dB in wind noise can be expected, depending on the SPL at the time, wind velocity, and the frequency of the sound pickup. Windscreens may be used with any type of microphone because they vary in their size and shape. Figure 1.162 shows a

A. Blimp-type windscreen.

B. Furry cover to surround the windscreen in A.

FIGURE 1.161   Blimp-type windscreen for an interference tube microphone. Courtesy Sennheiser Electronic Corporation.

windscreen produced by Shure employing a special type of polyurethane foam. This material has little effect on the high-frequency response of the microphone because of its porous nature. Standard styrofoam is not satisfactory for windscreen construction because of its homogeneous nature.

A cross-sectional view of a windscreen employing a wire frame covered with nylon crepe for mounting on a 1 inch diameter microphone is shown in Fig. 1.163. The effectiveness of this screen as measured by Dr. V. Brüel of Brüel and Kjaer is given in Fig. 1.164.

1.12.2.1 Wind Noise Reduction Figures for Rycote Windshielding Devices

Rycote has developed its own technique for measuring wind noise that uses real wind and a real time differential comparison. The technique compares the behavior of two microphones under identical conditions, one with a particular

FIGURE 1.162   Shure A81G polyurethane windscreen. Courtesy Shure Incorporated.

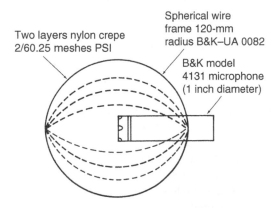

FIGURE 1.163   Typical silk-covered windscreen and microphone. Courtesy B and K Technical Review.

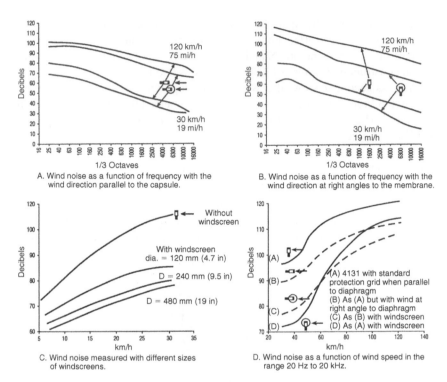

A. Wind noise as a function of frequency with the wind direction parallel to the capsule.

B. Wind noise as a function of frequency with the wind direction at right angles to the membrane.

C. Wind noise measured with different sizes of windscreens.

D. Wind noise as a function of wind speed in the range 20 Hz to 20 kHz.

FIGURE 1.164   The effectiveness of the windscreen shown in Fig. 1.163. Courtesy B and K Technical Review.

wind noise reduction device fitted and the other without, and produces a statistical curve of the result corrected for response and gain variations.

Figure 1.165 is a Sennheiser MKH60 microphone—a representative short rifle microphone—without any low-frequency attenuation in a wideband (20 Hz–20 kHz) test rig.

When a wind noise reduction device is fitted, its effect on the audio response is a constant factor—if it causes some loss of high-frequency, it will do it at all times. However, the amount it reduces wind noise depends on how hard the wind is blowing. If there is a flat calm it will have no beneficial effect and the result will be a degradation of the audio performance of the microphone. However, in a strong gale a small deviation from a perfect flat response may be insignificant for a >30 dB reduction in wind noise. Wind noise is in the low-frequency spectrum. For a naked Sennheiser MKH60, the wind-produced energy is almost entirely below 800 Hz, rising to a peak of 40 dB at about 45 Hz. It is the effect of a shield at these lower frequencies that is most important. Cavity windshields inevitably produce a slight decrease in low-frequency response in directional microphones but this is not usually noticeable. Basket types have

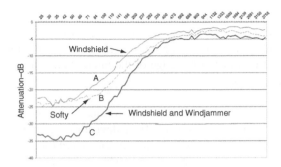

FIGURE 1.165 Wind noise reduction options for a Sennheiser MKH60 microphone under real wind conditions. Courtesy Rycote Microphone Windshields LTD.

very little effect on high-frequency. Fur coverings, while having a major effect in reducing low-frequency noise, will also attenuate some high-frequency.

Adding the low-frequency attenuation available on many microphones or mixers (which is usually necessary to prevent infrasonic overload and handling noise when handholding or booming a microphone) may give extra wind noise reduction improvements of >10 dB at the cost of some low-frequency signal loss.

The standard (basket) windshield shows up to 25 dB wind noise attenuation at 35 Hz while giving almost no signal attenuation, Fig. 1.165.

The Softy Windshield is a slip-on open cell foam with an integral fitted fur cover. The Softy reduces wind noise and protects the microphone. It is the standard worldwide in TV. A Softy attenuates the wind noise about 24 dB, Fig. 1.165B.

Adding a Windjammer (furry cover) to the basket windshield will give an improvement of about 10 dB at low-frequency to −35 dB, Fig. 1.165C. The attenuation of the Windjammer is approximately 5 dB at frequencies above 6 kHz although this will increase if it is damp or the fur is allowed to get matted. Overall this combination gives the best performance of wideband wind noise reduction against signal attenuation. To determine the correct windscreen for microphones of various manufacturers, go to www.microphone-data.com.

Pop protection is best appreciated when close-talking and explosive breath sounds are particularly bothersome. These explosive breath sounds are commonly produced when saying words involving P and T sounds. The phrase *explosive breath sound* is somewhat of a misnomer since these sounds, without amplification, are normally inaudible to a listener.[15]

The electrical output from the microphone is actually the transient microphone response to this low-velocity, high-pressure, pulse-type wave front. The P and T sounds are projected in different directions and can be shown by saying the letters P and T while holding your hand about 3 inches (7.6 cm) in front of your mouth. Note that the T sound is felt at a considerable distance below the P sound.

For most microphones, pop output varies with distance between the source and microphone, reaching a peak at about 3 inches (7.6 cm). Also the worst angle of incidence for most microphones is about 45° to the microphone and for a glancing contact just at the edge of the microphone along a path parallel to the longitudinal axis.

sE Dual Pro Pop Filter. An interesting pop filter is shown in Fig. 1.166. The sE Dual Pro Pop pop screen is a two-filter device to suit vocal performances. The device has a strong gooseneck with both a standard fabric membrane and a pro metal pop shield on a hinge mechanism. They can be used separately or both simultaneously depending on the application.

In an emergency, pop filters can be as simple as two wire-mesh screens treated with flocking material to create an acoustic resistance.

FIGURE 1.166   sE Dual Pro Pop pop screen. Courtesy sE Electronics.

1.12.2.2 Reflexion Filter

The Reflexion Filter by sE Electonics is used to isolate a microphone from room noises hitting it from unwanted directions, Fig. 1.167.

The reflexion filter has six main layers. The first layer is punched aluminum, which diffuses the sound waves as they pass through it to a layer of absorptive wool. The sound waves next hit a layer of aluminum foil, which helps dissipate energy and break up the lower frequency waveforms. From there they hit an air space kept open by rods passing through the various layers.

Next the waves hit an air space that acts as an acoustic barrier. The sound waves pass to another layer of wool and then through an outer, punched, aluminum wall that further serves to absorb and then diffuse the remaining acoustic energy.

The various layers both absorb and diffuse the sound waves hitting them, so progressively less of the original source acoustic energy passes through each

FIGURE 1.167 Reflexion Filter. Courtesy sE Electronics.

layer, reducing the amount of energy hitting surfaces so less of the original source is reflected back as unwanted room ambience to the microphone. The Reflexion Filter also reduces reflected sound from reaching the back and sides of the microphone. The system only changes the microphone output by a maximum of 1 dB, mostly below 500 Hz.

The stand assembly comprises a mic stand clamp fitting, which attaches to both the Reflexion Filter and any standard fitting shock mount.

1.12.3 Shock Mounts

Shock mounts are used to eliminate noise from being transmitted to the microphone, usually from the floor or table.

Microphones are very much like an accelerometer in detecting vibrations hitting the microphone case. Shock mount suspensions allow a microphone to stay still while the support moves.

Suspensions all use a springy arrangement that allows the microphone to be displaced and then exerts a restoring force to return it to the rest point. It will inevitably overshoot and bounce around, but the system should be damped to minimize this.

As frequency lowers, the displacement wavelength increases so the suspension has to move farther to do the job. For any particular mass of microphone and compliance (wobbliness) of suspension, there is a frequency at which resonance occurs. At this point the suspension amplifies movement rather than suppresses it. The system starts to isolate properly at about three times the resonant frequency.

The microphone diaphragm is the most sensitive along the Z-axis to disturbances. Therefore the ideal suspensions are most compliant along the Z-axis, but should give firmer control on the horizontal (X) and vertical (Y) axes to stop the mic slopping around, Fig. 1.170.

Suspension Compliance. Diaphragm and so-called donut suspensions can work well, but tend to have acoustically solid structures that affect the microphone's polar response. Silicone rubber bands, shock-cord cat's cradles, and metal springs are thinner and more acoustically transparent but struggle to maintain a low tension, which creates a low resonant frequency, while at the same time providing good X-Y control and reliable damping. The restraining force also rises very steeply with displacement, which limits low-frequency performance.

Shock mounts may be the type shown in Fig. 1.168. This microphone shock mount, a Shure A53M, mounts on a standard ⅝ inch – 27 thread and reduces

FIGURE 1.168   Shure A53M shock mount. Courtesy Shure Incorporated.

mechanical and vibration noises by more than 20 dB. Because of its design, this shock mount can be used on a floor or table stand, hung from a boom, or used as a low-profile stand to place the microphone cartridge close to a surface such as a floor. The shock mount in Fig. 1.169 is designed to be used with the Shure A89SM shotgun microphone.

Shock mounts are designed to resonate at a frequency at least 2½ times lower than the lowest frequency of the microphone.[21] The goal is simple but there are practical limitations. The resonant frequency (f_n) of a mechanical system can be computed from

$$f_n = \frac{1}{2\pi}\sqrt{\frac{Kg}{w}} \tag{1.32}$$

where,
K is the spring rate of the isolator,
g is the acceleration due to gravity,
w is the load.

A microphone shock-mount load is almost completely determined by the weight of the microphone. To obtain a low-resonant frequency, the spring rate or stiffness must be as low as possible; however, it must be able to support the microphone without too much sag and be effective in any position the microphone may be used.

The Rycote lyre webs rely primarily on their shape to give different performance on each axis. Typically, a 100 g force will barely move a microphone 1 mm along the (up and down) Y-axis, whereas it will move about four times that on the (sideways) X-axis. In the critical Z-axis, it will move almost ten times as far, Fig. 1.170.

FIGURE 1.169 Shure A89SM shock mount for a shotgun microphone. Courtesy Shure Incorporated.

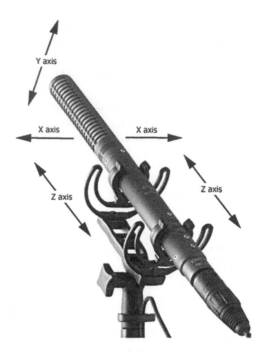

FIGURE 1.170   A Rycote lyre-type microphone suspension (shock mount) system. Courtesy Rycote Microphone Windshields LTD.

With a very low inherent tension the resonant frequency can be very low, and the Z displacement can be vast. Even with small-mass compact microphones, a resonance of <8 Hz is possible, which means that microphones can be well isolated across almost their entire frequency range.

Damping has to be added to metal spring suspensions, and although integral to rubber band versions, is not very easy to control. With the lyre webs damping can be selected almost independently by choosing a suitable plastic. The Hytrel that Rycote uses not only damps smoothly but maintains its characteristics even down to arctic temperatures. It also has a shape memory that allows it to be tied in eye-watering knots without developing a permanent set—or snapping!

Most suspension systems are difficult to scale. Springs and elastic bands become thin and fragile, and the range of softness for rubber and foam is limited. However, this does not apply to lyre webs. The tiny InVision suspensions, which are visually unobtrusive, isolate compact and similar sized microphones down to <30 Hz, yet are tough enough to be dropped on the floor without risk. Figure 1.171 shows the actual measured performance of the transfer function for a Schoeps CCM4 microphone being shaken with pink noise in an InVision mount. Trace A shows the output from the microphone with the shaker operating but not touching the mic, revealing the inherent coupling through air and the

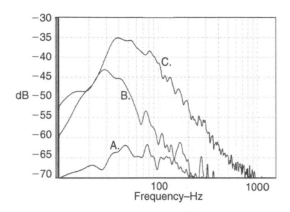

FIGURE 1.171   Effectiveness of an InVision mount. Courtesy Rycote Microphone Windshields LTD.

building itself. Trace B is with the shaker directly coupled to the microphone body to reveal the actual level of vibration input. Finally, the trace C shows the microphone's output with the shaker knocking the bar of the mount, thus demonstrating the effectiveness of the suspension.

To determine the correct suspension systems for microphones of various manufacturers, go to www.microphone-data.com.

1.12.4 Stands and Booms

Microphones are mounted on microphone floor stands or table stands to place the microphone in front of the sound source. The floor stands are usually adjustable between 32 and 65 inches (0.8–1.6 m) and incorporate a ⅝ inch – 27 thread for mounting the microphone holder or shock mount. They normally have a heavy base or three widespread legs for stability.

The table stands are 6–8 inches (15–20 cm) high and often incorporate a shock mount and an on–off switch, as shown in Fig. 1.172.

Small booms, which are mounted on the standard microphone floor stand, are normally used to put the microphone in a place where it is difficult to reach with a floor stand, Fig. 1.173. They are also useful when micing from above the source. Combination booms and stands are often on wheels or flat tripod legs and adjustable from 60–90 inches (1.5–2.3 m) vertically and 90–110 inches (2.3–2.8 m) horizontally, Fig. 1.174.

It is important that the boom and/or microphone stand be easily adjusted and that the clutch/brake system has a positive lock. Better microphone stands incorporate a piston-type air suspension system for effortless height adjustment and microphone protection.

Large booms, as used in television and motion-picture sound stages, are motorized and often include a stage for the microphone sound person.

FIGURE 1.172   Electro-Voice table microphone stand with push-to-talk switch. Courtesy Electro-Voice, Inc.

FIGURE 1.173   Atlas BB-44 microphone boom. Courtesy Atlas Sound.

1.12.5 Attenuators and Equalizers

Attenuators, equalizers, and special devices from Electro-Voice, Shure, and others are available to reduce the microphone output level or shape the response to roll off the low or high end, increase the 3–5 kHz articulation region, or reverse polarity. These units normally have standard input and output male and female XLR or ¼ inch phone plug connectors. Attenuators are also available to be installed between the capacitor capsule and the condenser microphone electronics to eliminate overload from high-level sources.

FIGURE 1.174   Adjustable microphone stand/boom. Courtesy Atlas Sound.

1.13 MICROPHONE TECHNIQUES

Micing is more of an art than a science. Therefore there is no one way to position a microphone for good recording. It is subjective and at the control of the engineer. The discussions of microphone placement in the following sections are only suggestions or the ideas of one engineer.

The quality of the reproduction can be greatly influenced by the position of a microphone in relation to the sound source. When only one microphone and one sound source are involved, this positioning is fairly straightforward: the closer the microphone, the more the direct sound will dominate over the reverberant sound. Except in an anechoic chamber, there will always be a certain amount of reflected sound present in the microphone output. This results from sound bouncing off boundaries such as the floor, ceiling, walls, and objects of significant proportions located in the area of the microphone. At a certain distance from the sound source, the amount of reflected sound will exceed the amount of the direct sound. The microphone is then said to be in the reverberant, or far, field. The effect is to make the acoustic environment (usually a room) more evident to the listener than would be the case with close micing (microphone in the near field).

The proper position of the microphone depends on the effect desired. Close micing produces a highly present, up-front sound, with little of the acoustic environment evident, whereas distant micing produces a more spacious sound with the room characteristics becoming very obvious. A close microphone position may not accurately reproduce the sound of the source, and equalization may be required to achieve a sound similar to the natural sound. If the room acoustics are not suited to the sound reproduction desired, a distant microphone position may produce an unpleasant or unintelligible result. The correct choice requires the engineer to choose the appropriate microphone position for the sound desired. A microphone placed an inch from a snare drum will produce an up-front, bigger-than-life sound, which could be appropriate for a modern rock recording but might be totally inappropriate for a jazz or big band recording. Distant micing of the snare drum could produce a powerful effect, in any kind of music, since the contribution of a good room might be important to the music.

It is rare that there is just one microphone and one sound source. Modern recording often requires the use of multiple microphones. Microphone placement then becomes more complicated, because as the microphone is moved farther from its intended source, more of the other sources will be picked up as well. No instrument is a point source, and there are different characteristic sounds emanating from various places on the instrument (i.e., a flute has vastly different sounds coming from the open end, the body of the flute, or the mouthpiece). Most instruments have complex directional characteristics that vary from note to note. Even instruments of the same make and model can sound quite different from one another.

Whenever there is more than one microphone receiving sound from a single source, a problem of time and phase differences can become audible. This problem can have a major effect on the frequency response, presence, and clarity of the recording. The result for spaced microphones can be a comb filter effect, which will tend to reduce presence, upset the natural balance of various notes and overtones, and disturb localization of the source. In an extreme case, certain notes may be attenuated to inaudibility. In practice, the contribution of room reflections, pickup by other microphones, and intrinsic instrument imbalances may mask many of these effects.

Multitrack recording generally requires the engineer to isolate instruments so that only the intended source is recorded on each track. Sometimes this is simple because the track is being overdubbed and only that one instrument is in the studio. At the other extreme, an entire ensemble may be playing at once, yet the situation may require that all instruments be totally isolated on the tape tracks so that they can be individually mixed, processed, or even replaced with no effect on the other instruments. The latter requires very careful microphone choice and placement and/or the use of isolation booths for some troublesome instruments. If the musical balance is good in the room, the job is fairly simple. But if there are obviously incompatible instruments playing simultaneously (e.g., heavy drums versus a finger-picked acoustic guitar), isolation solely through microphone technique becomes next to impossible.

1.13.1 Stereo Micing Techniques

Modern recording practice often employs multiple microphones, each feeding a separate track of a multitrack tape machine. Sound reinforcement practice usually requires good isolation of the various sound sources. In either case, the end result is a composite of a number of monaural sources, which are often placed in the stereo image with pan pots. This practice is not the same as true stereo recording, which can provide a sense of depth and realism unachievable with panned mono sources. It requires greater effort for superior results; a good acoustic environment is essential.

There are a number of stereophonic recording techniques available to the engineer. The simplest requires two microphones, often omnidirectional types, spaced apart by a distance ranging from several feet to more than 30 ft (9 m), Fig. 1.175. The spacing depends on the size of the sound source, the size of the room, and the effect desired. A broad source like an orchestra will require a wider spacing than a small source such as a single voice or instrument. If the microphones are too far apart, a hole in the middle of the stereo image will result, since the sound produced in the center of the stage will be too far from either microphone. When placed too closely together, a mono result will be obtained. When the spacing is comparable to the wavelength of the sound, phase cancellations may result (comb filters), which will destroy the monaural compatibility of the recording. The best spacing seems to be from 10–40 ft (3–12 m).

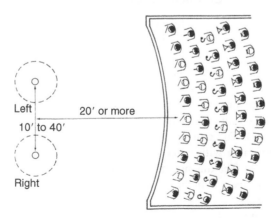

FIGURE 1.175   Spaced omnidirectional microphones for stereo recording.

Experimentation is necessary since every situation will be different. Needless to say, good monitoring is required; stereo headphones will not generally reveal defects evident on good monitor loudspeakers. A method of summing the two channels to mono is essential for testing compatibility.

Variations on the spaced microphone technique involve using bidirectional or unidirectional microphones, which may be helpful when the room characteristics are not perfect for the material being performed. Adding a microphone in the center, fed to both left and right channels (*fill* microphones), and combinations of spaced micing and other techniques might be required.

1.13.2 Microphone Choice

Every microphone type has certain characteristics. These characteristics must be taken into account when choosing a microphone for a specific application. Some of the factors to be considered are general type (condenser, moving coil dynamic, ribbon); directional pattern (omni-, bi-, or unidirectional); and specific microphone traits (bright, bassy, dull, presence peak, and so on).

Also, the susceptibility of the microphone to overload or its tendency to overload the associated preamplifier must be considered. The off-axis frequency response can have a large effect on the sound of a microphone in a particular application. Certain microphones may exhibit unusual traits that may make them more, or less, suitable for a certain application. For example, the design of the grille may have a major effect on the sound of a microphone when recording closely micing vocals.

Some of these characteristics can be inferred from the microphone specifications (i.e., frequency response, overload point, directional pattern—both on- and off-axis). Other characteristics are not as easy to measure or visualize, and experience and experimentation are necessary to make an intelligent choice.

1.13.3 Microphone Characteristics

There are many criteria used to judge the suitability of a microphone for a particular application—some are quite subjective. Frequency response is one obvious characteristic, distortion is another. The ability of a microphone to accurately translate waveforms into electrical signals is vital for good reproduction. Generally, the less massive the internal parts that must be moved by the sound pressure, the more accurate the reproduction, especially the reproduction of waveforms with steep leading edges and/or rapid level changes (e.g., percussive sounds). The condenser microphone has the lowest mass (only a thin plastic diaphragm with a very thin coating of metal must be moved by the sound pressure). The diaphragm and coil in the dynamic microphone have considerably more mass than the condenser diaphragm. The ribbon in a ribbon microphone has relatively low mass and is somewhere between the condenser and the dynamic microphone.

It would seem that the condenser microphone would always be the best choice, but other factors must be considered. Condenser microphones are generally less rugged than dynamic ones, and since they are usually more expensive, the decision to place a valuable microphone in a position where it could be hit or knocked over must be weighed against the possible benefit of improved sound. Also, condenser microphones contain internal active electronics, which can be overloaded by high sound levels. Many condenser microphones contain switchable or insertional pads, but long before the overload distortion becomes apparent, clipping of the transient peaks may muddy the sound in a subtle way.

Ribbon microphones are somewhat fragile. They can be especially vulnerable to blasts of air that can occur when closely micing vocals, inside a bass drum, or even when a door is slammed in an airtight studio.

In each type of microphone, there are many other factors that can affect the sound. The design of the mounting for the microphone components, the internal obstacles in the sound path, and the effect of the body of the microphone, all can have a major effect on the ultimate sound reproduction.

1.13.3.1 Directional Pattern

It might at first seem that the unidirectional microphone would be the universal choice for all applications, since picking up the intended source is the goal. It is true that unidirectional microphones (see Section 1.2.3) have the greatest application, but there are situations that require the use of omnidirectional microphones, which are designed to pickup sound from all directions as nearly equally as possible (see Section 1.2.1), or bidirectional microphones, which are sensitive to the front and back, but insensitive to the sides (see Section 1.2.2). But it is possible, in some situations, to obtain greater rejection of unwanted sound with an omni- or bidirectional microphone than would be possible with a unidirectional pattern.

Unidirectional and bidirectional microphones often exhibit a proximity effect, in which the response to lower frequencies (generally below 150 Hz) is emphasized when the microphone is placed close to the sound source (Section 1.2.3). Close may be a couple of inches or a couple of feet, depending on the microphone. Various designs have been developed to minimize or eliminate this effect. A switchable high-pass filter may be included on the microphone to roll off the bass in close micing positions. Proximity effect must be considered when choosing and placing a microphone. Sometimes the effect can be used to advantage (i.e., when additional bass response is desirable, perhaps on a snare drum or on certain vocals). But often the proximity effect emphasizes the (unrelated) tendency of some sound sources to sound more bassy when close mic'ed.

Directional microphones do not have the same frequency response off-axis as they do on-axis. This can cause increased apparent sound leakage from other sources, tonal aberrations of the reproduced sound, or unexpected phase cancellations. For example, many directional microphones exhibit less directionality at both higher frequencies and lower frequencies. If such a microphone were used to close mic a snare drum, the amount of pickup of the nearby bass drum and cymbals might be excessive.

1.13.4 Specific Micing Techniques

There are probably as many methods of using microphones as there are engineers. Contrary to popular opinion, there does not seem to be any special microphone or magical technique for recording any particular sound. What is right is what sounds best. The following discussion is merely a review of some common techniques widely employed and likely to work well in many circumstances.

1.13.4.1 Musicians

The first requirement for obtaining a good sound from any instrument is a superior player. An experienced studio musician can make almost any studio or engineer sound good. Unfortunately, the engineer usually has very little to say about the musicians who are hired for the session. When inexperienced players record, they may often expect to be made to sound like whoever their idols may be. They probably don't want to know that their idol spent the last 10 years or more learning how to use the studio, and they may be likely to blame the engineer for their inability to play properly for recording. There isn't much that can be done in such a circumstance.

1.13.4.2 Drums

Studios involved in music recording are more often judged by their drum sound than by anything else. It is true that much contemporary music relies heavily on drums and that getting the best possible sound is a goal worth pursuing. There are any number of ways to record drums, but the most commonly used technique utilizes close micing.

Just as the musician is a vital element in obtaining a good sound, the drums themselves must be in good condition and properly tuned to obtain their best sound. The type of drum head used will have a major effect on the sound.

Micing Each Drum. A micing arrangement that is almost standardized requires the use of one microphone on each drum, Fig. 1.176. In addition, one or more microphones may be suspended over the drum set to pickup either an overall sound or primarily cymbals. How closely each microphone is placed depends on several factors: how tight a sound is required, which in turn is related to the relative liveness and character of the room; what isolation problems might exist, in terms of various drums leaking into other drum microphones and leakage

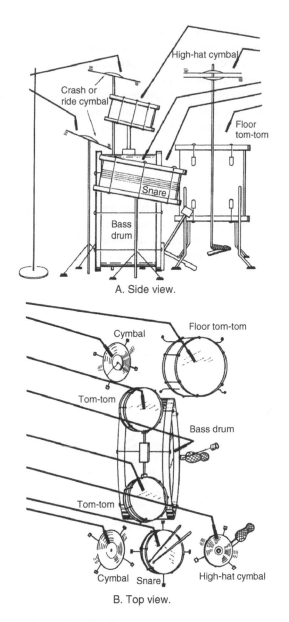

FIGURE 1.176 Close drum micing (detail).

from other instruments in the room; how dangerous it may be to place an expensive and fragile microphone in a position of possible destruction by an overly enthusiastic or inaccurate drummer; and whether the microphone and/or console can take the level produced without distortion.

Above versus Underneath Micing. Individual drums can be miced either from above or below, Fig. 1.177. The two positions will usually have vastly different sounds. If the sound is appropriate, the underneath position may be preferable if isolation is a problem.

When miced from above, microphones are commonly positioned at an angle to the drum head and near the edge of the drum. Seemingly minor changes in position can have a major effect on the sound, especially with some microphones.

Bass Drum. For recording, bass drums usually have only the beaten head, which is not to say that bass drums with both heads cannot be recorded, however. For some music, the use of both heads is preferable. In the single-head configuration, the usual microphone placement is within the shell of the drum, with the microphone aimed toward the beater, Fig. 1.178. Experimentation is required, however. Closer or farther distances, off-axis microphone positions, or even placement on the opposite side of the head may result in the desired sound.

Tom-Tom Micing. Tom-toms, too, often use only the top head. This facilitates underneath micing. In micing any drum, it is probable that simultaneous top and bottom micing will result in difficulty due to phase discrepancies. The use of phase-reversal switches at the board and minor position adjustment may be required.

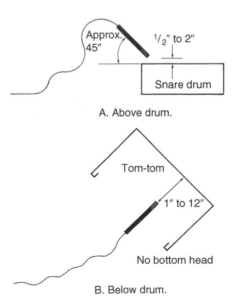

FIGURE 1.177   Bass drum micing.

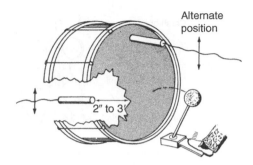

FIGURE 1.178   Bass drum micing.

Cymbals. The high hat and cymbals can be mic'ed from above or below, but the above position is more commonly used. Overhead microphones are often positioned above the entire drum set, usually as a stereo pair. How high they are set will depend on the effect desired; a relatively high placement will provide more of an overall drum sound, with more room characteristics than a closer position.

It is not unusual to pickup sufficient cymbals, or even excessive cymbals, from just the other drum microphones without the overhead microphones even being on. The amount of cymbal leakage will be determined mostly by the drummer's technique and balance, with the room characteristics also being a factor.

Other Drum Micing Techniques. Close micing every drum is only one method. Another is to use relatively distant microphones to pickup an overall drum sound, Fig. 1.179. This, of course, results in much more room sound and possible leakage from other instruments. It also requires that the drummer play all the drums, and particularly the cymbals, in the proper balance. The engineer has much less control of the sound. This approach will not be successful in poor rooms, nor with drummers who do not correctly balance their various drums and cymbals. But in a good room, with a good drummer, the sound can be quite natural and often very powerful. A common technique is to use two overhead microphones, placed in such a way as to capture the natural sound and balance of the drum set. Some experimentation will be required to find the proper placement. Usually a separate bass drum microphone is used as well, to give the bass drum better definition and more punch.

1.13.4.3 Piano

Pianos are often recorded in stereo and can add width and a greater sense of space if done properly. Multiple microphones spread all over the sounding board may seem like an ideal way to pickup the full piano sound, but this procedure can lead to a very artificial and distant sound when heard in mono.

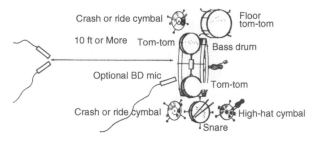

Crash or ride cymbal

10 ft or More   Tom-tom

Floor tom-tom

Bass drum

Optional BD mic

Tom-tom

Crash or ride cymbal

High-hat cymbal

Snare

A. Coincident microphones.

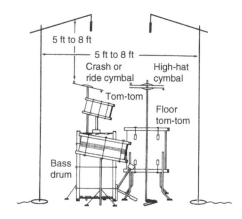

5 ft to 8 ft

5 ft to 8 ft

Crash or ride cymbal

High-hat cymbal

Tom-tom

Floor tom-tom

Bass drum

B. Overhead spaced microphones.

FIGURE 1.179   Distant drum micing.

First, be certain that a stereo piano is really desirable. In multitrack recording, are there sufficient tracks available? And is the piano sound required to be so big? A mono piano can often have more punch and might be a better choice.

For a mono track, one microphone is usually all that is needed. For stereo, a pair of adjacent directional microphones will probably suffice. In micing either a grand or upright piano, keep in mind that the sounding board and not the hammers and strings is the source of most of the sound. With a good piano, there may be surprisingly little difference in the sound picked up from various areas of the sounding board. A commonly mic'ed point is where the bass and treble strings cross, Fig. 1.180. Variations such as micing from beneath (or in the case of an upright, in back of) the sounding board, inserting microphones into the circular holes in the harp, or using various types of piano pickups can all be tried. Each piano is different, and each player will also have a large effect on the sound, so a variety of techniques should be tried.

The PZ Microphones can be used in recording piano. They can be placed on the inside of the piano lid and the lid closed for improved isolation.

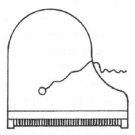

A. Single microphone or coincident pair (typical).

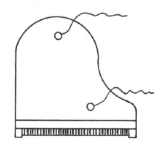

B. Spaced microphone (typical).

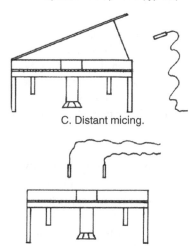

C. Distant micing.

D. Overhead micing (piano lid removed).

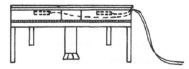

E. PZM micing–lid closed
(placement similar to B).

FIGURE 1.180 Piano micing.

As in all percussive instruments, the peak level produced by a piano can be far greater than the level shown on the volume unit (VU) meter. Peaks 20 dB above the meter reading are common. Since just about everybody knows what a piano sounds like, and since the instrument is so frequently featured in musical pieces, any distortion will be very obvious to the listener. Even a distortion that only occurs on the peaks can be evident as a dulling of the piano attack, a kind of audio blurriness. The peaks can really strain the dynamic range of microphones, preamps, and storage medium. If condenser microphones are used, be sure the pads are switched on even if the level seems moderate. Also, some engineers routinely record piano at a somewhat lower than normal level to avoid tape saturation.

Obtaining satisfactory isolation while still getting a good sound can be a problem with the piano. Isolation can be achieved with a booth, of course, but careful micing and some baffling can often work almost as well. One technique used in many studios is to place the microphone in the piano and then close the lid as much as possible. Often the short-stick position of the lid works well. Then carpeting or other dense, heavy, absorbent material is draped over the piano. With a good arrangement of other instruments and reasonably balanced volumes, very little leakage should exist. Another technique requires that microphones be mounted inside the piano, usually suspended from the lid, in such a way that the lid can be completely closed. The PZM type of microphone is particularly well suited for this approach.

Of course, a much better sound is obtained with the lid open and with perhaps a little more distance between the sounding board and the microphones. Sometimes removing the lid and suspending the microphones above the piano work well. (Most pianos have pins in the hinges that can be easily removed for this purpose.) Fairly distant micing may also sound good.

1.13.4.4 Vocals

A single vocal, either speaking or singing, is usually recorded with one microphone placed within 2 ft (0.6 m) of the mouth. For popular music, it is common to have the singer very close to the microphone; in a recording of a classical singing voice, a greater distance is appropriate, even up to several feet may be used if the room acoustics permit. Speakers at a lectern usually are 1–2 ft (0.3–0.6 m) from the microphone.

1.13.4.5 Singers

Although vocals could be recorded in stereo, with any of the techniques previously described, it is customary to record the voice in mono. It is basically a point source, with little directional information. In a superior acoustic environment, such as a good concert hall, natural reverberation may be mixed in with additional microphones. But most often artificial reverb is added. It can be stereo and add considerable depth and width to the voice.

Condenser microphones, placed very close to the mouth, are the usual choice in the studio. A pop filter will be necessary for all but the most careful singers. This prevents explosive sounds from being produced when the vocalist sings a word containing Ps or other hard consonant sounds. It is important to remember that the output level of the microphone will adhere to the inverse square law, which states that if the distance from the source to the microphone is doubled, the level will be reduced to one-quarter. Experienced vocalists are well aware of this phenomenon and may even use it to obtain certain effects. The inexperienced or inattentive singer will probably require electronic processing (i.e., limiting) to obtain a satisfactory performance. This problem is further complicated by the trend toward mixing vocals quite low in the musical track and relying on processing to maintain intelligibility.

In the studio, it is often necessary to provide an acoustic environment less reverberant than normal for the recording of vocals. Cutting down on reverberation could be accomplished with a separate vocal booth with highly sound-absorbent surfaces, or it can be obtained by placing absorbent baffles around the singer and microphone in the studio, Fig. 1.181. On the other hand, it may sometimes be necessary to emphasize the reverberation for a special effect by distant micing or by mixing in another microphone placed some distance away.

Proximity effect can be a problem with vocals. Many microphones have provision for a bass roll-off, which can be used to correct this deficiency. This approach is often superior to using equalization in the control room, especially if a limiter is used before the equalizer (the limiter would respond to the emphasized bass and thus not accurately track the vocal intensity). Some singers prefer the effect obtained from proximity, using the bass boost in their performance to emphasize certain words or phrases.

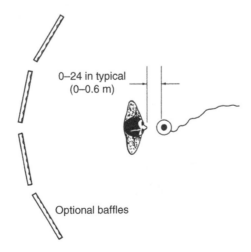

0–24 in typical
(0–0.6 m)

Optional baffles

FIGURE 1.181   Vocal micing.

In a live performance, large studio condenser microphones would be inappropriate. With their suspensions and pop filters and the large microphone stand required, they would obscure the singer's face. What is needed is a relatively small, rugged microphone that can be handheld if desired. Although there are a number of condenser microphones that can be used this way, the usual choice is a compact dynamic microphone with built-in pop filters, integral shock mounting, and switchable bass roll-off.

Good directionality is required of a live performance microphone. The usual practice of providing the singer with a stage monitor loudspeaker, usually placed within a few feet of the microphone, requires good rejection of sound from off axis to minimize the possibility of feedback and reduce the degradation of the sound from the vocal microphone picking up the monitor's reproduction of the other instruments and voices. Some microphones designed for live work have their direction of minimum sensitivity oriented toward the direction where the most unwanted sound would come from (i.e., not directly off the back of the microphone, but at some intermediate angle).

1.13.4.6 Group Vocals

A vocal group could consist of two singers or a chorus of several hundred. For a small group (less than eight), a single microphone with an omnidirectional pattern placed in the center of a circle of vocalists often works well, Fig. 1.182. This microphone arrangement requires that the singers achieve a proper balance of voices in the studio. The final balance can be fine tuned by having the necessary voices move closer to or farther from the microphone. If the singers are relatively close to the microphone (2 ft or less), then their positions become more critical. A small change in position can have a major effect on the blend.

For stereo, the group could be divided into two circles, each with its own omnidirectional microphone in the center. Two bidirectional microphones, oriented at 90° to one another and placed one above the other, could be used to

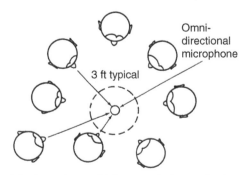

Adjust spacing to obtain proper balance of voices

FIGURE 1.182   Group vocals micing monaural.

obtain a stereo omnidirectional recording when placed in the circle of vocalists, Fig. 1.183.

Whenever omnidirectional microphones are used, the room becomes more apparent in the recording than it would with unidirectional microphones. This effect must be considered when recording group vocals in this manner.

If greater presence is required (or less room sound) or if the balance must be controlled by the engineer for some reason, individual microphones could be used for each singer; however, this method has obvious practical limitations if the group is large. It also requires more set-up and balancing time, puts a musical burden on the recording personnel, and might have a disappointing result if lack of isolation creates phase problems when mixing the multiple microphones.

For really large groups, techniques similar to those described for string sections might be employed.

Typically, group vocals will be recorded as an overdub on a previously recorded musical track, requiring the vocalists to wear headphones. With a number of singers wearing headphones (which could be turned up quite loud) standing next to an omnidirectional microphone, a significant amount of leakage from the headphone mix is possible. This leakage from the headphone mix can become even more of a problem if one or more of the singers prefer to remove one side of the headphones from his or her ear in order to better hear their own voice and/or the blend of the other voices. Background vocals are often by nature relatively quiet parts requiring higher than normal gain on the microphone channel. All these factors can combine to degrade the entire recording seriously.

Solutions might be to use as low a headphone level as possible, have the singers sing as loudly as is appropriate for the part, turn the microphone off when the vocalists are not singing, or use a noise gate to do this automatically. In a really severe situation, the solution might be to use individual directional microphones.

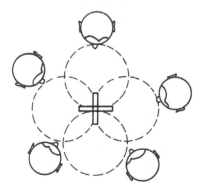

Bidirectional microphones at 90° to one another

FIGURE 1.183 Group vocals micing stereo.

1.13.4.7 Lectern Microphones

For redundancy, two or more microphones are often provided on a lectern. Only one should be active at a time, or phase cancellations can result. Often two microphones are arranged on opposite sides of the lectern, angled in toward the talker. The goal is satisfactory pickup as the speaker moves from side to side. This arrangement can cause serious phase cancellation problems because of the spacing (usually a couple of feet) resulting in feedback problems since the normal frequency response has been disturbed through the comb filter effect. A better arrangement places the two microphones in the coincident configuration as close together as possible and angled toward opposite sides of the lectern, Fig. 1.184. The outputs can be summed with no phase problems. The angle between the microphones may be changed from the normal 90° if necessary to obtain proper coverage.

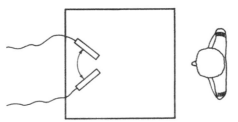

FIGURE 1.184   Lectern microphones for increased coverage pattern.

1.13.4.8 Strings

Although strings could be close miced, this approach usually results in an unnatural sound. Distant micing is more appropriate but puts a greater demand on the room acoustics. Obtaining a good string sound really requires a good room of considerable size.

A string quartet might sound fine recorded in a relatively small studio ($2200\,\text{ft}^3$ or $62\,\text{m}^3$), but a large string section needs more volume. Not only will a larger room accommodate more players, but the microphone placement will also be simpler and the results will be closer to the actual sound of the section.

Each instrument could have a microphone, and this would give the mixer complete control of the balance of all the strings. But unless a great deal of time is available to obtain the proper balance, this approach is not cost effective when recording highly paid musicians. It does not guarantee the best results, either.

At the opposite extreme, a single microphone placed at a point determined to provide the best overall balance and sound could be a simple and quick way to get good results, Fig. 1.185. This placement works well if the engineer is familiar with the room and can rapidly duplicate a setup that has been successful in the past. A coincident pair can provide the same sound if stereo is required.

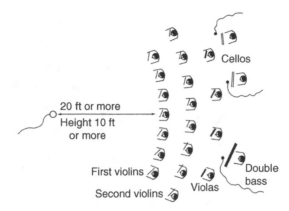

FIGURE 1.185   Single microphone (or coincident stereo) for string section recording. Optional microphones are shown for cello and double bass.

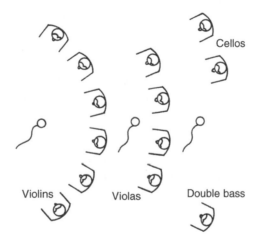

FIGURE 1.186   String micing by section.

Another technique is to mic the ensemble in sections, Fig. 1.186, providing, for example, a single microphone for the first violins, another for the second violins, another for the violas, and so on. Cello and double bass often have microphones to pick them up individually in this type of setup.

It is also possible to set up microphones above each row of players, or above each two rows. This method is often used in conjunction with the single overall microphone.

In a practice session, Fig. 1.187, the setup is often a composite of all of these techniques: a single coincident pair at a distant point (perhaps 15–20 ft [5–6 m]

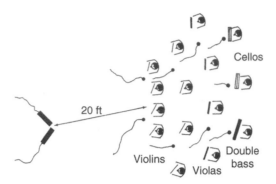

FIGURE 1.187   Typical composite technique for string micing.

from the first row, and up as high as practical in the room); a set of microphones over each section (one microphone for every two players, up above the space required for their bows and slightly in front of the instrument); and individual microphones for the cellos and basses (a foot or two in front of the instrument, opposite the F holes). At the start of the session, the overall microphone would first be monitored to determine what, if any, balance problems exist. If time permits, the overall microphone position might be changed to obtain a better balance. If the desired balance cannot be obtained with the single microphone, the necessary individual section microphones may be brought into the mix. In many practically sized rooms, it is not possible to obtain a good balance of the near strings (usually violins) and the far strings (cello and bass). Careful use of the section microphones can correct this.

Since good high-frequency and transient response is required to reproduce the string section sound, condenser microphones are the most frequent types used for string recording.

1.13.4.9 Horns

In the recording world, horns are any brass instrument: trumpets, trombones, saxophones, and so on. Modern recording of popular music usually requires close micing of individual instruments, Fig. 1.188. Since many horns are capable of producing very high sound-pressure levels (as high as 130 dB), it is important to choose microphones that will not be overloaded by this close placement. Also, pads may be required to prevent overloading the mixer preamplifier or saturating an input transformer.

Condenser microphones are often used to pickup horns, but ribbon and dynamic types may also give good results.

It is important to remember that the sound produced by these instruments does not come entirely from the bell; this is particularly true of saxophones. Although the instrument output may be loudest at the bell, the contribution of the various other parts of the horn cannot be ignored. The microphone position

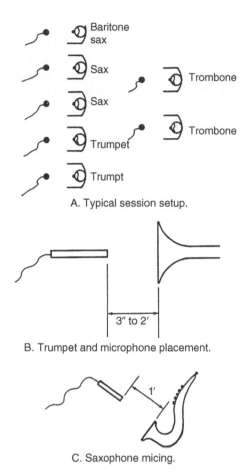

A. Typical session setup.

B. Trumpet and microphone placement.

C. Saxophone micing.

FIGURE 1.188   Horn and microphone placement.

is often a compromise between the presence of very close placement, the better tonality of a slightly greater microphone distance, the leakage from other instruments as the microphone distance is increased, and the degree of room contribution desired in the finished recording. Depending on the effect desired, 6 inches to a couple of feet may be appropriate.

1.13.4.10 Woodwinds

Instruments like the oboe, flute, bassoon, clarinet, and their variations cannot generally have microphones placed too closely and still retain their character. In popular music they are often mic'ed individually at a distance of one to several feet, which is generally not far enough to provide a true sound, but the result is often acceptable—or even desirable—for compatibility with other instruments in the song.

Condenser or ribbon microphones are the usual choice. Most woodwinds tend to sound most natural when mic'ed from about 3 ft (1 m) away, with the microphone directed toward the middle of the instrument, or perhaps pointing slightly toward the bell or end of the instrument. Low placement, even on the floor with a PZ Microphone, tends to sound better than high placement.

For classical recording, a more distant pickup is necessary. A woodwind ensemble might be successfully recorded using the techniques described previously for string sections.

1.13.4.11 Electric Instruments

In this category are all instruments designed to be reproduced through amplifiers and loudspeakers. Electric guitar; electric bass; various synthesizer, organ, and other electronic keyboards; and acoustic instruments with attached microphones or pickups designed for amplification all fall into this category.

Generally, these instruments require microphone placement with the associated amplifier/loudspeaker combination. However, another technique is possible and in many cases preferable—that is, the direct recording of the instrument. Since most of these instruments produce a microphone-level high-impedance unbalanced output, all that is required in most cases is a high-quality transformer, providing the match between the instrument and the low-impedance balanced inputs of most mixers. Various direct boxes are available, some with active electronics to provide the required impedance transformation. Almost all provide an output to drive the instrument amplifier as well as the mixer, and most have a ground switch to select the grounding configuration with the least noise.

In many situations, the instrument and its amplifier constitute a system. The amplifier, which may contain loudspeakers or may be connected to a separate loudspeaker system, may have a major effect on the sound of the instrument. Taking a direct feed may result in a totally unnatural sound.

Micing the instrument amplifier may seem simple, but often the cabinet contains several loudspeakers. These may be identical loudspeakers or separate drivers for various frequency ranges. A single close microphone may not provide the proper balance. Even in systems with identical loudspeakers, careless microphone placement may result in phase discrepancies producing a distant and/or colored sound. Two solutions are practicable: either give the microphone a more distant placement, far enough to be equidistant from all the loudspeakers, or position it very close, to pickup only one loudspeaker, Fig. 1.189.

Systems with multiple drivers for different frequency ranges will have to be mic'ed from far enough away that the various drivers are properly balanced. Although it may be possible to mic the individual drivers and mix them for the proper balance, this approach is more prone to error.

Distant micing is often desired, especially for an electric guitar. Naturally, the character of the room must be appropriate.

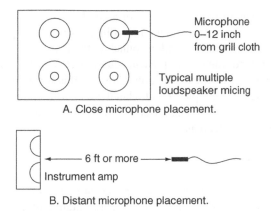

FIGURE 1.189   Micing of electric instrument loudspeakers.

Often a combination of the direct and mic'ed sound is used. This combination can be effective, but the phase relationship between the two sources will be arbitrary, which can cause severe coloration of the sound. The tonal balance will change unpredictably as the ratio of direct and mic'ed sound changes. This change usually precludes any gain riding of the individual inputs. A phase reversal switch can sometimes be used to optimize the gross phasing between the two inputs.

Instruments like synthesizers or other electronic keyboards generally should be recorded directly. The sound of these instruments is usually not augmented by the addition of a musical instrument amplifier. There are exceptions, however, and the choice of technique depends on the effect desired—perhaps the limited frequency response and soft distortion of a tube-type amplifier is appropriate.

1.13.4.12 Percussion

The most common percussion instrument is the drum kit. Other percussion instruments, such as congas, tympani, handclaps, tambourines, timbales, wood blocks, claves, or maracas, etc., require care in micing due to the extreme levels encountered. It is not uncommon to have levels of +10dBm and more (open circuit) on the output of a condenser microphone when placed close to a percussion instrument or a piano.

Such levels can be very demanding of microphone electronics, in the case of condenser microphones and the associated mixer. The use of internal microphone pads is essential. Additional padding may be necessary between the microphone output and the mixer input.

The correct micing procedure for percussion instruments depends on the effect desired. A distant micing position is often justified when the sound of the room reverberation adds to the effectiveness of the instrument. Tambourine and handclaps often benefit from the sound of a good room. The resultant sense of space can produce better depth in the recording, and/or the explosive nature of a large, live room can add tremendous punch to the part.

On the other hand, the highly present sound of close micing might be more appropriate in another musical situation. Close, in this sense, might range from fractions of an inch to a couple of feet. Handheld instruments, like claves, must be played at a uniform distance from the microphone, which becomes more critical as the distance decreases.

1.13.5 Conclusion

It is important to remember that there is never only one way to position microphones. The techniques presented here are representative of the methods widely used in the recording and sound-reinforcement industries, but such practices have evolved over many years. Some are traditional; however, there may be better ways. Using the procedures outlined will result in reasonably accurate reproduction, or commercial reproduction as it applies to mainstream music recording. Since sound reproduction can be a creative endeavor, experimentation may yield new techniques. The exact reproduction of the original sound may not be the goal. Perhaps the engineer is attempting to obtain a previously unheard sound or effect. When the luxury of experimentation is available, the engineer may well use the time to pioneer new techniques that can supplement or even replace existing procedures.

ACKNOWLEDGMENTS

Thanks to Michael Pettersen, Shure Incorporated, for his assistance in updating and correcting this chapter.

REFERENCES

1. A.M. Wiggins, "Unidirectional Microphone Utilizing a Variable Distance between the Front and Back of the Diaphragm," *ASA*, Vol. 26, no. 5, September 1954. (Parts copied with permission.)
2. R. Schulein, Shure Incorporated, "Development Consideration of a Versatile Professional Unidirectional Microphone," *Journal of the AES*, Vol. 18, no. 1, p. 44, February 1970. (Parts copied with permission.)
3. W. R. Bevan, R. B. Schulein, and C. E. Seeler, Shure Incorporated, "Design of a Studio-Quality Condenser Microphone Using Electret Technology," *J. Audio Eng. Soc.*, vol. 26, no. 12, p. 947, December 1978.
4. H. Tremaine, *Audio Cyclopedia*, Indianapolis, IN: Howard W. Sams & Co., Inc., 1969, pp. 148–150.
5. Ibid., pp. 150–151, 152.
6. A. P. G. Peterson and E. E. Gross, Jr., *Handbook of Noise Measurements*, General Radio Co., p. 33.

7. Don and Carolyn Davis, *Sound System Engineering*, Indianapolis, IN: Howard W. Sams & Co., Inc., 1975.

8. Henning Gerlach, "Stereo Sound Recording with Shotgun Microphones," *Sennheiser News*. (Parts copied with permission.)

9. Yuri Shulman, "Reducing Off-Axis Comb-Filter Effects in Highly Directional Microphones," *J. Audio Eng. Soc.*, Vol. 35, no. 6, June 1987.

10. Y. Ishigaki, M. Yamamoto, K. Totsuka, and N. Miyaji, Victor Company of Japan, Ltd., "Zoom Microphone," *AES preprint 1718 (A-7)*. (Parts copied with permission.)

11. Manfred Hibbing, "XY and MS Microphone Techniques in Comparison," *Sennheiser News*. (Parts copied with permission.)

12. Wesley L. Dooley and Ronald D. Streicher, "M-S Stereo: A Powerful Technique for Working in Stereo," *AES preprint* 1792 (J-6).

13. David A. Ross, "A Practical Approach to Applying the MS Stereo Microphone," Shure Brothers, Inc.

14. C.P. Repka, "A Guide to Coincident Mikes," *Audio Magazine*, November 1978.

15. Klaus Genult and Hans W. Gierlich, "Investigation of the Correlation between Objective Noise Measurement and Subjective Classification," HEAD Acoustics GmbH, Aachen, Federal Republic of Germany.

16. Dr. Klaus Genuit and Wade R. Bray, "Binaural Recording for Headphones and Speakers," *Audio Magazine*, December 1989.

17. H.W. Gierlich and K. Genuit, "Processing Artificial-head Recordings," *AES preprint 2460 (G-5)*.

18. Don and Carolyn Davis, "In-the-Ear Recording and Pinna Acoustic Response Playback," *AES preprint 2874 (W 3/5-E)*.

19. Joe Ciaudelli, "Large Multi-channel Wireless Microphone Systems," *Sennheiser Electronic*. (Parts copied with permission.)

20. "RF-Transmission Technique, Wireless Microphones and Reporters Sets," *Sennheiser Electronic*.

21. G.W. Plice, "Microphone Accessory Shock Mount for Stand or Boom Use," *J. Audio Eng. Soc.*, Vol. 19, no. 2, February 1971. (Parts copied with permission.)

BIBLIOGRAPHY

J. Eargle, *The Microphone Book*, Focal Press, Burlington, MA, 2004.

Microphones: An Anthology of Articles on Microphones from the Pages of the Journal of the Audio, New York: Audio Engineering Society (AES), 1979.

Don Davis and Eugene Patronis, Jr., *Sound System Engineering*, Focal Press, Burlington, MA, 2006.

David Miles Huber, *Microphone Manual*, Howard W. Sams & Co., Indianapolis, IN, 1988.

Loudspeakers

Jay Mitchell

2.1 INTRODUCTION

A loudspeaker is a device that converts electrical energy into acoustic energy (electroacoustic transducer), or more generally, a system consisting of one or more such devices. Loudspeakers are present in our daily lives to such an extent that, in most modern societies, one is in almost constant contact with them. From the time the speaker in our clock radio wakes us in the morning until we turn off the television before we go to bed at night, we encounter loudspeakers almost constantly. Even our computers have loudspeakers.

A general treatment of loudspeakers, including their history and design considerations, in order to fit within a single chapter of a book such as this, is limited to providing an overview of the subject rather than an in-depth treatment of design and theoretical considerations. We will touch on as many of the relevant areas as available space permits, while providing references for the reader who is interested in further study. This chapter may serve as an overview of the subject for end users and audio enthusiasts and as a guide to further study for those interested in performing loudspeaker design work themselves.

2.1.1 Uses of Loudspeakers

Even though there is a very wide range of applications for loudspeakers, they may be thought of as serving some combination of four primary purposes:

1. Communication.
2. Sound reinforcement.
3. Sound production.
4. Sound reproduction.

While there are common requirements for all of these uses, each one also imposes its own demands on loudspeaker attributes. In a given application, it is possible that more than one of these purposes must be served by a single loudspeaker. In such cases, the suitability of the loudspeaker for one or more of its uses may be compromised in order to facilitate others.

Communication. Ranging from intercom systems in offices and schools to radio communications systems for the space shuttle, voice communication systems make our everyday lives safer and more convenient. The first practical loudspeaker was in the earpiece of the original telephone. Since that time, loudspeakers have been an integral part of voice communication systems, from intercom systems to satellite-based telephone and conferencing systems.

Sound Reinforcement. In numerous situations involving public speaking and musical performance before audiences in halls, auditoriums, amphitheaters, and arenas, the sound created by the voices and/or musical instruments is not of sufficient loudness to be heard or understood satisfactorily by everyone present. In such situations, a sound reinforcement system can provide the acoustic gain required to overcome this deficiency.

Sound Production. There are a number of subcategories of this type of loudspeaker usage. Perhaps the most readily recognizable is the use of amplification as an integral part of certain musical instruments—e.g., electric guitar, bass, and keyboards. Other examples include emergency warning and sonar systems. Loudspeaker characteristics may be very highly specialized when they are used as part of a sound production system, and loudspeakers optimized for this type of use are often not well suited to other uses.

Sound Reproduction. Playback of recorded music, motion picture soundtracks, and videotape requires a sound reproduction system. Almost every home in the United States has one or more sound reproduction systems. Movie theaters and recording studios also require sound reproduction systems. One of the author's past design projects was a loudspeaker system for use in an international chain of large-screen specialty theaters.

2.1.2 Loudspeaker Components

It is useful to identify the component parts (or subsystems) of a loudspeaker for individual examination and analysis. For purposes of this chapter, the components of a loudspeaker are:

1. Transducer.
2. Radiator.
3. Enclosure.
4. Crossover.

We will examine various forms of each of these components in the sections that follow. Their interactions with each other within a loudspeaker will be discussed. We will also present concepts of loudspeaker performance characterization and an overview of electroacoustic models. The reader is encouraged to pursue the subject matter that is presented here through the references provided in the bibliography. The design and analysis of loudspeakers is a multidisciplinary field, incorporating elements of music, physics, electrical and mechanical engineering, and instrumentation. The individual subject areas are challenging and fascinating in and of themselves, and their convergence in the field of loudspeaker design results in one of the most complex combinations of art and science that has ever existed.

2.2 TRANSDUCER TYPES

There are a number of ways in which electrical energy can be converted into acoustic energy. Of all the possibilities for carrying out this function, a relative few have become dominant in practical loudspeakers: electrodynamic, electrostatic, and piezoelectric. In general, an electroacoustic transducer contains three elements: motor, diaphragm, and suspension. The motor converts electrical energy into mechanical (motional) energy and the diaphragm converts mechanical energy into acoustic energy (vibration of the transmission medium, usually air). A suspension supports the diaphragm, allows it to move in an appropriately constrained fashion, exerts a restoring force proportional to displacement from its equilibrium position, and provides a damping force proportional to the velocity of motion that serves to prevent the diaphragm from oscillating in an undesired manner.

2.2.1 Electrodynamic Transducers

The most common type of transducer used in loudspeakers is the electrodynamic driver. In this type of transducer, a time-varying current passing through a conductive coil suspended in a time-invariant magnetic field creates a force on the coil and the parts to which it is attached. This force causes the parts to vibrate and to radiate sound.

There are a number of viable implementations of electrodynamic transducers. By far the most common is the cone driver. In a cone driver, a cone-shaped diaphragm is suspended at its outer periphery by a structure called a *surround* and (usually) near its center by a *spider*. The *motor* consists of a permanent magnet assembly that concentrates the magnetic field in an annular gap, in which is placed a voice coil that is attached to the center of the cone via a cylindrical coil former. An electrical signal is applied to the voice coil, and the current in the *voice coil* interacts with the magnetic field in the gap to create a time-varying force that vibrates the diaphragm. Figure 2.1 shows a typical cone driver. The most commonly used magnetic material is ferrite, or ceramic. Other magnetic materials used in loudspeakers include aluminum/nickel/cobalt (*alnico*) and neodymium/iron/boron (*neodymium* or *neo*). The magnet structure is typically held together with an anaerobic thermoset adhesive. Some loudspeakers are assembled with bolts through the magnet. In this case, stainless steel or brass screws must be used, so as not to magnetically short the top plate to the backplate. A rear cover may or may not be used. A vent through the pole piece may be provided. It serves to prevent the addition of a spring constant due to

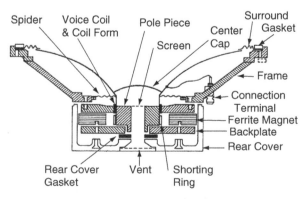

FIGURE 2.1   Typical woofer parts identification. Courtesy Yamaha International Corp.

the small air cavity under the center cap (dust cover) and to reduce turbulence-induced noise due to pumping effects in the magnet gap.

2.2.2 Diaphragm Types

The most common direct-radiation device is the cylindrical voice coil–driven paper cone. The cheapest cone to make is the folded cone, which is cut from a sheet of paper, rolled, and bonded at the seam. A more expensive and difficult to make cone is the molded-paper cone. These are one piece, molded by straining a slurry of water and paper pulp through a strainer mold in the shape of the desired end product. The formed wet mat of pulp is then pressed and baked to remove residual moisture, bearing a dry, strong one-piece cone, free of joints. Ribs and concentric rings are sometimes molded into the cone, and the cones can be formed with straight or curved sides of varying depth. These are all available from suppliers of cones.

While most mathematical models of a direct radiator assume a rigid piston, in practice this is impossible to achieve. In some cases, diaphragm rigidity is intentionally reduced in order to produce specific desired behavior. Two examples involving a controlled breakup are shown in Figs. 2.2 and 2.3. The whizzer cone in Fig. 2.2 is intended to radiate high frequencies as the larger cone decouples from the motor. The Biflex principle, as popularized by Altec Lansing in the 1950s, is shown in Fig. 2.3. The inner cone is attached via a compliant element at A to the large outer cone in hopes of decoupling the outer cone at high frequencies.

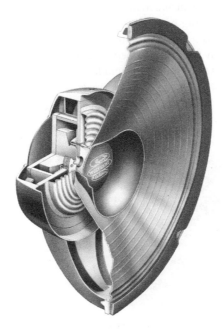

FIGURE 2.2 Loudspeaker incorporating a whizzer cone.

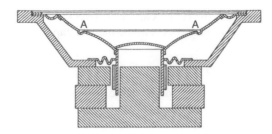

FIGURE 2.3   Loudspeaker illustrating decoupling center cone. (From U.S. Patent 4,146,756.)

Damping dope is applied to the coupling connection in an attempt to smooth the decoupling transition frequency response. While whizzer cones are still in use in some inexpensive ceiling speakers, there are at this time no devices similar to the Biflex on the market.

In addition to felted paper, a number of newer materials have found use in cone-type low- and medium-frequency loudspeakers. A variety of plastics have been used, the most popular being polypropylene and bextrene. The KEF Company introduced a composite aluminum-skinned foam-core sandwich cone. Community Professional Loudspeaker's M4 compression midrange similarly uses a carbon fiber/epoxy composite diaphragm. Adamson Acoustics in Canada uses a Kevlar fabric resin-bonded diaphragm for the midrange driver. Mitsubishi Electric (Japan) introduced a studio monitor, which used cone woofers fabricated from a honeycomb core/carbon fiber skin composite.

In a loudspeaker with an alnico magnet, the magnet is directly under the pole piece (as opposed to being between the top and backplates), and the outside of the magnet structure is a cast iron return from the bottom of the magnet to the top plate. Venting may be accomplished via a hole covered with open wire mesh in the center dome. Other methods include a uniformly porous dome with no magnet vent, Fig. 2.4.

FIGURE 2.4   Alnico magnet woofer–Altec 515-8LF. Courtesy of Altec Lansing Corp.

2.2.3 Suspension Methods

The suspension of a cone driver comprises two distinct components: the surround and the spider. The surround is attached to the periphery of the diaphragm or cone and is itself attached to the support structure (the basket in the case of a cone driver). The spider is attached to the voice coil former (or to the cone in the vicinity of the former) and is also attached to the basket on its periphery. Because they affect cabinet sealing, surrounds are designed to be nonporous. Surrounds and spiders both contribute to the damping of the motion of the diaphragm. The most popular surround construction is heat-formed, open-weave, resin-impregnated linen with formed-in convolutions and sealed with damping dope. Other surrounds are made of foam or butyl rubber formed in a half-roll. On some loudspeakers, a viscoelastic (never-drying) dope is applied to the surround.

Spiders are usually made of a heat-formed, open weave, resin-impregnated cloth that is formed into convolutions. They are usually not treated with a sealing material (dope). The unsealed fabric is needed for venting, since the air beneath the spiders can otherwise be trapped. This also tends to damp the spider. An early method of making porous spiders was to die-cut them from solid phenolic-impregnated linen sheet stock. The spider is not required to seal the edge of the cone to its enclosure as is the surround. In a typical cone driver, the spider contributes the majority of the stiffness in the suspension.

2.2.4 Mechanical Construction

The Peavey Black Widow bass drivers are unusual in that they have a streamlined magnet structure, called focused-field geometry, Fig. 2.5. It employs a magnetic circuit that has smoothly flowing flux lines, as might be intuitively preferred for a fluid flow channel. Other manufacturers have adopted similar approaches to magnet design. An added benefit of this approach is minimization of weight.

JBL ferrite-magnet drivers have symmetric field geometry. Figure 2.6 shows the top plate configuration, which makes the magnetic leakage flux at the top and

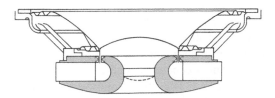

FIGURE 2.5 Peavey focused-field geometry magnet structure with one-piece backplate/pole piece forging. Courtesy Peavey Electronic Corp.

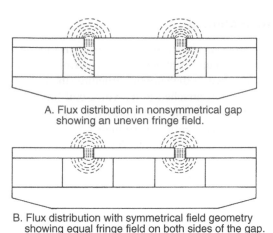

A. Flux distribution in nonsymmetrical gap
showing an uneven fringe field.

B. Flux distribution with symmetrical field geometry
showing equal fringe field on both sides of the gap.

FIGURE 2.6 JBL symmetric field geometry versus asymmetric design. Courtesy JBL.

bottom of the gap symmetric, thereby, according to the manufacturer, reducing magnetic drive asymmetry and the resulting low-frequency distortion.

Another form of electrodynamic transducer is the dome radiator. Most commonly used for high frequencies, dome radiators have the advantages of compactness and predictability of acoustic behavior. Domes can be made from linen, impregnated phenolic fabric, Mylar™, paper, aluminum, titanium, beryllium, and composites such as carbon fiber/epoxy. Soft dome tweeters have been in widespread use for a number of years. Some of this popularity may be due to the fact that there is no abrupt transition from piston radiation to breakup. Instead, most of the radiation from a soft dome comes from the region immediately adjacent to the voice coil, making it function as a ring radiator.

Several flexible diaphragms have been used on magnetic drivers, all sharing the same basic construction: etched aluminum conductors on Mylar™ film. These are operated in various magnetic field configurations to produce sound. One of the earliest of these is the Magneplanar® loudspeaker, which consisted of an entire field of magnets over which the diaphragm conductor was mounted. Magneplanars are in the shape of large panels. The Heil high-frequency driver, used in systems manufactured by ESS, used direct radiators similar to Magneplanar® in that the voice coil was printed on Mylar™. The ESS-Heil unit, however, was corrugated, and the sound was produced by these vertical pleats moving open and closed, thereby squeezing air into radiated sound. An extension of this was used also by ESS in the Transar system, which used hollow spheres modulated by electromagnetically driven rods. Mitsubishi Electric (Japan) developed a printed conductor high-frequency device called the *leaf tweeter*, as shown in Fig. 2.7. The ribbon loudspeaker is the simplest and has excellent potential for good high-frequency response due to the fact that the diaphragm is the conductor. No extra diaphragm structure is used on the ribbon.

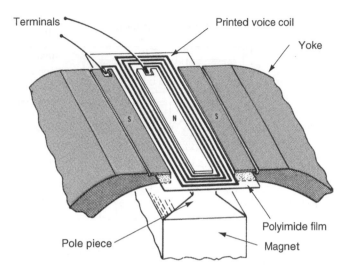

FIGURE 2.7   Technics Leaf Tweeter diaphragm detail. Courtesy Panasonic Industrial Corp.

2.3 COMPRESSION DRIVERS

One means of improving the performance of an electrodynamic transducer that will be used to drive a horn is to create a compression driver. In a compression driver, the diaphragm radiates into a compression chamber and its output is typically directed through a phasing plug to the driver's exit, which is attached to the throat of the horn.

The advantage of a compression driver is that relatively small diaphragm velocities are converted to larger particle velocities at the exit of the driver. The effect of this transformation is that less diaphragm excursion is required for a given acoustic power output. The tradeoffs for this coupling include possible increases in certain distortion components and the requirement for a horn. Compression drivers are not used as direct radiators.

The purpose of the phasing plug is to equalize path lengths from the diaphragm surface to the exit. To the extent that this is accomplished, the useful bandwidth of the driver will be extended upward in frequency.

Figure 2.8 is a cross-sectional view of a typical ceramic magnet wide-range compression driver using a dome diaphragm. The case construction is unusual and peculiar to this design by Yamaha. The phase plug is also a bit unusual; however, it is still of the circumferential-slit variety on the phase plug (dome) surface. The diaphragm is aluminum and is supported by a typical Bakelite™ or plastic support frame. The back cap has sound-absorbing material inside to discourage interfering air resonances in the cap.

Figure 2.9 shows a 2 inch throat JBL driver, model 2440, using an alnico magnet. The phase plug is more typical than that in Fig. 2.8, using more straight through circumferential slits. The JBL plug is made of cast Bakelite.

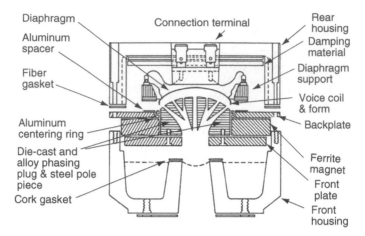

Diaphragm

Aluminum
spacer

Fiber
gasket

Aluminum
centering ring

Die-cast and
alloy phasing
plug & steel pole
piece

Cork gasket

Connection terminal

Rear
housing

Damping
material

Diaphragm
support

Voice coil
& form

Backplate

Ferrite
magnet

Front
plate

Front
housing

FIGURE 2.8 Typical ceramic-magnet compression driver. Courtesy Yamaha International Corp.

FIGURE 2.9 JBL 2440 2 inch throat alnico compression driver. Courtesy JBL/UREI.

Figure 2.10 shows another alnico driver, the 1 inch Altec 802/808. The 802 uses an all-aluminum diaphragm with tangential surround coupled via the phase plug and expanding throat section to a 1 inch diameter exit. The 808 is identical to the 802 in all respects except for the diaphragm. From left to right in the exploded view are the pot, the alnico magnet slug that fits under the pole piece, which is mounted on a radial-slit tangerine phase plug. This unusual design is made from glass fiber–filled plastic and is bonded to the pole piece. Above this (to the right) is the ring that centers the pole piece in the air gap via the top plate. It is nonmagnetic (brass), and the holes provide a mechanical load on the voice coil, which affects response and distortion. Next are the diaphragm assembly and rear cap.

FIGURE 2.10   Altec 802/808 alnico 1 inch throat compression driver with tangerine radial phase plug. Courtesy Altec Lansing Corp.

The preceding three examples are representative of dome diaphragm compression driver design practices, both in concept and in practical implementation.

A large number of variations exist in the art including a wide variety of suspension shapes and materials. Domes are made from aluminum, titanium, and beryllium. Yamaha International Corp. makes its suspension out of stamped beryllium-copper cantilever fingers, and JBL and Ramsa stiffen their suspensions with rhombic and diamond patterns, respectively; this is actually a redistribution of diaphragm breakup resonances.

Midrange compression drivers are useful where there is a requirement to supply very high levels of acoustic power with low distortion. Community Professional Loudspeakers, M4 midrange driver is shown in Figs. 2.11 and 2.12. It is intended for use from 200 Hz to 2000 Hz. The diaphragm is approximately 7 inches in diameter. Originally it was fabricated from specially formed aluminum skins and a light, stiff foam core, about 0.090 inch thick. More recent versions had diaphragms made of a carbon fiber composite.

FIGURE 2.11   Community M4 4 inch throat midrange driver. Courtesy Community Professional Loudspeakers.

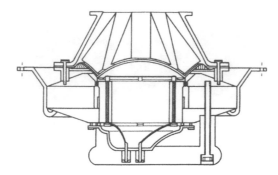

FIGURE 2.12   Community M4 cross-section view. Courtesy Community Professional Loudspeakers.

Another compression driver configuration is the screw-on driver. The University 7110XC (explosion proof) is shown in Fig. 2.13. This type of unit is often used on a reentrant horn in public address systems. Throat diameter is usually ¾ inch. Diaphragms are most often made of phenolic resin-impregnated domes with integral convoluted suspensions. Voice coils are usually round copper wire.

FIGURE 2.13   University 7110XC ¾ inch throat public address driver.

2.4 ELECTROSTATIC TRANSDUCERS

Electrostatic transducers make use of the fact that two static electrical charges placed at a distance from each other will experience a force directed along a line between them. The force is attractive if the charges have the opposite sign (positive and negative) and repulsive if the charges have the same sign (positive and positive or negative and negative). In practical loudspeaker designs, the forces are attractive, due to the complementary nature of the charge transfer from the amplifier output to the speaker plates. The magnitude of the force is inversely proportional to the distance between the charges and directly proportional to the magnitude of the charges.

A typical electrostatic loudspeaker consists of a diaphragm made of two pieces of metallic foil separated by a sheet of dielectric, or nonconductive, material. By itself, the application of a pure ac signal (i.e., one with no dc component) to an electrostatic loudspeaker would cause attractive forces for both positive- and negative-going signal excursions, since the induced charges are opposite in both cases. This would create a frequency-doubled signal containing extremely high levels of harmonic distortion.

For this reason, a dc polarizing voltage is applied to the foil diaphragms, maintaining a steady attraction between them. The audio (ac) signal is superimposed on this dc offset, modulating the attractive force. In response to this modulated force, the diaphragms move opposite (toward or away from) each other. The upper limit on the amplitude of the allowed signal voltage is then equal to half the polarization voltage. This arrangement is the basis of all modern electrostatic loudspeakers. The result is an acceptably low level of harmonic distortion, as long as variations in the distance between plates or the diaphragms are minimized. The movement of the foil diaphragms generates sound waves. The diaphragms produce equal acoustical power radiated in opposite directions. This set of characteristics defines a dipole radiator.

It is asserted by the designers of electrostatic loudspeakers that they overcome certain basic disadvantages of cone-type loudspeakers, particularly with respect to the propagation of acoustic energy at the high frequencies. Cone-type loudspeakers are driven by a voice coil that is attached to a relatively small portion of the total diaphragm area, and they do not behave as pistons at higher frequencies. Because the electrostatic loudspeaker has a diaphragm that is driven uniformly across its surface, breakup is said to be eliminated. Additionally, the diaphragm can have low mass compared to the air load on the diaphragm. This enhances high-frequency and transient response.

Electrostatic loudspeakers may be constructed in several different ways. Two of the most common construction types are:

1. Stretching the diaphragm between supports around its periphery and leaving an air gap between the diaphragm and two stationary electrodes, Fig. 2.14.
2. Using an inert diaphragm that is supported by a large number of tiny elements disposed across the entire surfaces of the two electrodes. These elements act as spacers to hold the diaphragm in the center between the electrodes, Fig. 2.15.

In the latter type of loudspeaker, the diaphragm is a thin sheet of plastic on which has been deposited a very thin layer of conductive material. It is supported by multiple small elastic elements that hold the diaphragm in place but permit it to follow audio-signal waveforms. The electrodes on each side of the diaphragm are acoustically transparent to avoid pressure effects from trapped air as well as

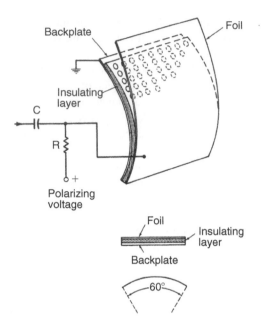

FIGURE 2.14   Electrostatic- or capacitor-type loudspeaker.

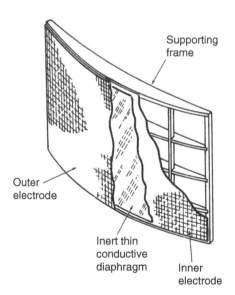

FIGURE 2.15   Cutaway view showing the internal construction of an electrostatic loudspeaker.

to permit acoustic energy to propagate away from the diaphragm. This type of construction permits the diaphragm to be of arbitrary size. The performance per unit area is the same for any area of the diaphragm. The actual loudspeaker is a thin surface curved in the horizontal, forming a section of a cylinder.

A surface that is large with respect to wavelength becomes increasingly directional at high frequencies.

Since an electrostatic loudspeaker is designed to couple directly with the acoustic resistance of air, the mass of the diaphragm is quite small and can be neglected with little effect on the accuracy of predictive models. The velocity of the diaphragm is directly proportional to the electrostatic force applied, except as altered by the stiffness of the diaphragm suspension. Measurements indicate that for a constant voltage applied to the electrodes, the acoustic response is uniform (flat) to well beyond the range of human hearing.

Output at low frequencies is limited by the maximum linear amplitude of the diaphragm motion, which is determined by spacing between the diaphragms and damping in the suspension. The maximum power output from an electrostatic loudspeaker of a given diaphragm area is determined by the strength of the electrostatic field that can be produced between the diaphragm and the electrodes.

For linear operation of an electrostatic loudspeaker, it is necessary to impose a dc polarizing voltage across the plates. To see why this is necessary, consider a sine wave applied to the loudspeaker: on the positive-going half of the wave, the movable plate is attracted to the fixed plate, with the peak attraction occurring at the positive voltage peak. This attraction returns to zero along with the input signal. However, on the negative half of the cycle, the plates are once again attracted to each other. The movable plate goes through the same action as it did on the positive-going part of the input signal. Thus, two pulses, both moving in the same direction, have been produced in response to a single cycle of the applied sine wave.

With a polarizing voltage applied, a steady electrostatic force is created between the plates, and the movable plate is attracted to the fixed plate.

Applying a sine wave (plus), the movable plate is attracted to the fixed plate beyond its position fixed by the polarizing voltage. Upon reversal of the sine wave, the electrostatic force between the plates is reduced, and the movable plate returns to its zero position. On the negative half of the cycle, the polarizing voltage is reduced because the sine wave is of opposite polarity, and the attracting force is decreased below the polarizing voltage value. Therefore, the plate moves away from the fixed plate, completing the cycle. In this manner, frequency doubling has been prevented, and the loudspeaker produces a sine wave output in response to a sine wave input. By connecting the plates for push-pull operation, as shown in Fig. 2.16, the signal is split between the two sides of the loudspeaker, and distortion is further reduced.

An electrostatic loudspeaker is seen by an amplifier as a capacitor with a value on the order of 0.0025 μF from electrode to electrode. Thus, the magnitude of the impedance presented by the loudspeaker to the output of the amplifier falls off at 6 dB per octave as the frequency is increased. This presents some problems for driving electrostatics, as many amplifiers are not designed to drive purely capacitive loads.

Because electrostatic loudspeakers are relatively large in area compared to cones, their directivity is high in comparison to cone systems. Various schemes have been used by designers of electrostatics to address this issue. The Quad

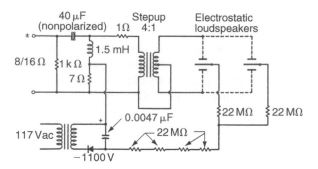

FIGURE 2.16   Typical coupling circuit and high-voltage power supply for electrostatic loudspeakers.

ESL63 is one example. Here, the diaphragm is broken into different regions for different frequency ranges, the smaller ones being used for higher frequencies, thereby making them wider in dispersion than a large single panel.

2.5 PIEZOELECTRIC LOUDSPEAKERS

Piezoelectricity, or pressure electricity, was discovered in the 1880s by the Curies. It is today a feasible motor drive mechanism for loudspeakers. In a piezoelectric material, a voltage applied to the material will result in a mechanical strain or deflection. The reverse is also true, and piezoelectric elements can be used in microphones. This characteristic is attractive for direct-drive units such as ultrasonic devices. For loudspeakers, however, some means must be applied to mechanically amplify the inherently low excursion so that a loudspeaker diaphragm may be driven properly.

One of the earliest discovered piezoelectric substances is Rochelle salt. Although Rochelle salt is still widely used, it suffers from poor mechanical strength, low temperature breakdown (55°C), and extreme sensitivity to humidity. Barium titanate is the first piezoceramic to be developed. Although it is not as electrically sensitive, it is still widely used, exhibiting many superior characteristics over Rochelle salt. The most widely used piezo material today is lead zirconate titanate, developed first in Japan in the 1950s. This material (PZT) is now highly refined and exhibits the best properties of any piezo material for loudspeaker use.

PZT material is formed by baking a ceramic slurry or clay into bars about 1 inch in diameter and then slicing the bars into thin wafers. Two wafers are bonded together in opposing polarity, with electrodes on their flat surfaces, forming a bimorph bender. As voltage is applied to the bender, deformation of the disc results in greater displacement at its center.

Early commercial attempts at the application of bimorph benders to loudspeaker cones involved a rectangular drive element anchored at three corners, allowing the fourth corner to drive the loudspeaker cone fore and aft. Other attempts used a cantilever structure anchored at one end with the loudspeaker cone mounted at the other. In 1965 when Motorola, Inc. first manufactured a piezoelectric loudspeaker, they used a length expander tube driving a horn-loaded cone directly. This device, like most piezoelectric loudspeakers made until that time, still lacked sufficient voltage sensitivity to be coupled directly to conventional systems without using an auxiliary step-up transformer.

The development of the circular bimorph using a corrugated center vane represented the next step forward in piezo loudspeaker technology. The action of the two disks working against each other, one expanding while the other contracts, functions as a mechanical transformer, giving impedance reductions of about 20:1. The basic operation is as follows: The driver dishes in and out; it pumps the cone fore and aft or, in the case of the horn, into a compression chamber that is then coupled to the throat of a horn via a slot and flared rib construction. The driver is allowed to hang free in space, working against its own inertia to pump the cone.

Further advancements in the state of the piezoelectric art came from Tamura and coworkers in their work on piezoelectric high-polymer films. This concept of a diaphragm possessing piezoelectric properties and thus coupling directly to the air without the use of any separate motor structure represents a substantial advancement toward the ideal acoustic transducer.

Another problem area in the development of the PZT loudspeaker was in the power-handling capability of the driver. The theoretical failure mode of a piezoelectric tweeter is the depolarizing of the driver through excessive drive level and/or high temperature. The Curie point (depolarizing temperature) of the PZT used here is above 150°C, and the depolarizing voltage is 10 V/25 µm thickness, or about 35 Vrms for the basic driver. These numbers describe a fairly impressive power-handling capability, but unfortunately one that is reached only in theory. In reality, under continuous high drive levels, mechanical stress on the surface of the ceramic wafers generates cracks in the microstructure that eventually penetrate the entire wafer. This is especially severe around the area where solder connections are made to the wafers, since the soldering operation tends to prestress the material at this point. The net result is that the 35 V maximum drive level is an intermittent specification, with the continuous drive level recommended at a 15 V maximum up to 20 kHz. For use above that frequency, it has been recommended that the level be reduced further through the addition of a series attenuation resistor so as to safeguard the ceramic element from absorbing excessive high-frequency power. Here again, when using larger, thinner ceramic wafers, these problems are further aggravated. Using this ceramic is an area for future development.

Figure 2.17 shows the Motorola KSN 1001A. Although Motorola manufactures a wide variety of other piezo-driven loudspeakers, the one illustrated here is the most widely used.

One near-optimal application of piezoelectric drive is underwater use. This is due to the excellent impedance match of the piezoelectric material to water via a waterproof barrier. Lubell Labs manufactures the underwater loudspeaker shown in Fig. 2.18. Although swimming pool loudspeakers using standard electromagnetic drivers are also available, the piezoelectric configuration is more

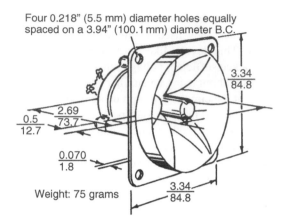

Four 0.218" (5.5 mm) diameter holes equally
spaced on a 3.94" (100.1 mm) diameter B.C.

$\dfrac{3.34}{84.8}$

$\dfrac{2.69}{73.7}$

$\dfrac{0.5}{12.7}$

$\dfrac{0.070}{1.8}$

Weight: 75 grams

$\dfrac{3.34}{84.8}$

FIGURE 2.17   Motorola KSN 1001A piezoelectric ultrahigh-frequency driver/horn. Courtesy
Motorola, Inc.

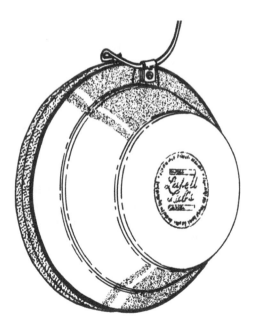

FIGURE 2.18   Lubell Labs underwater piezoelectric loudspeaker. Courtesy Lubell Laboratories, Inc.

efficient due to its mechanical impedance match to water. The loudspeaker is
fixed to the side of the pool and driven like a conventional loudspeaker. Lubell
Labs makes high-power arrays of these devices and a portable swim coach
system with a noisecanceling microphone for underwater communications in
various pool athletic events.

2.6 MOTOR DESIGN CONSIDERATIONS

The most common means of coupling amplifier output to the diaphragm in an electrodynamic transducer is via a cylindrical voice coil. This configuration is used on all magnetic cone loudspeakers and compression drivers. This is commonly known as a linear motor. The coil, made of round or rectangular wire (edgewound), is wound around a hollow cylinder called a *former*. Formers may be made of paper, plastic (e.g., Kapton polymer, Mylar™), or aluminum. The voice coil assembly is bonded to the diaphragm. Figure 2.19 shows the construction of a typical cone loudspeaker.

One novel approach to motor design involves printing or etching a conductor onto a thin sheet of Mylar™ (0.0005 inch) then folding it to produce a pleated diaphragm that is forced in the magnetic field. In another implementation, continuous lengths of wire are bonded to a large panel of Mylar™, which is operated over a field of bar magnets. The leaf tweeter is similar, etching a conductor field on Mylar™. They are identical in principle to Fig. 2.19 and are discussed more thoroughly elsewhere in this text. The ribbon loudspeaker, Fig. 2.20, is a special case in which the voice coil serves as both conductor and diaphragm.

FIGURE 2.19   Cross section of a typical cone loudspeaker showing construction (alnico magnet at center under pole piece). Courtesy JBL.

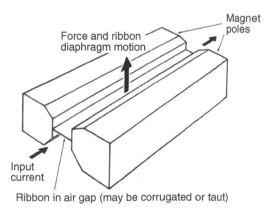

FIGURE 2.20   Ribbon loudspeaker.

One notable departure from conventional linear motor design is the Servo-Drive loudspeaker. This patented drive system uses a rotary servomotor that drives a woofer cone and suspension assemblies via a pulley-belt mechanism, alternately pushing and pulling the diaphragms in response to the input signal. Two opposing diaphragms are driven in a push-pull arrangement so as to yield a balanced axial force on the drive mechanism. The motor is configured so that it presents a typical impedance load to an amplifier. SDL (for servo-drive loudspeakers) speakers come in a variety of sizes and power capacities, but typically they are in the form of low-frequency horns. Figure 2.21 shows the mechanism employed to translate rotational motion of the servomotor to the linear motion needed to drive the opposing diaphragm assemblies. The opposing reinforced blastomeric belt mechanisms are used in the rotation-to-linear conversion, and the result is noiseless and free of slip. The opposing diaphragms drive the throats of conventional wood-fabricated folded bass horns. The positions of the diaphragms are shown in Fig. 2.22. Figure 2.23 shows the positioning of the servo-driven diaphragms in a typical folded bass horn.

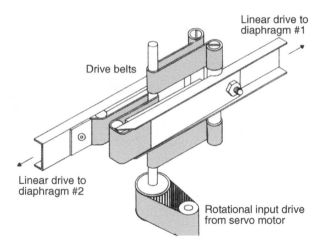

FIGURE 2.21   Belt drive system of the SDL loudspeaker.

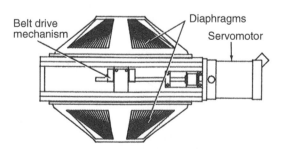

FIGURE 2.22   Position of SDL belt drive and opposing diaphragms.

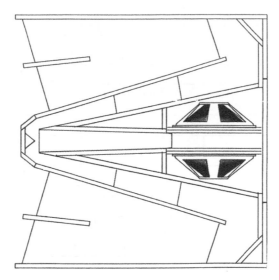

FIGURE 2.23 Position of the SDL diaphragms on a folded horn.

2.6.1 Output Limitations

The maximum usable output of an electromagnetic loudspeaker is a function of a number of parameters, including diaphragm displacement, heat transfer, sound quality (maximum acceptable nonlinearity), and/or wear life due to fatigue of moving parts.

There are two fundamental limitations on a magnetic driver, a displacement limit and a thermal limit. Displacement limits may be caused by either mechanical or electrical factors. Mechanical displacement limiting occurs when a moving part contacts a stationary one or when a suspension element is made unacceptably nonlinear (either temporarily or permanently) by deformation beyond its design range. Electrical displacement limiting occurs when the motor is operated outside its range of linear travel. This is a function of the length of the windings on the voice coil and the thickness of the plates that form the magnet gap. Figure 2.24 shows three typical voice coil configurations: equal length, overhung, and underhung coils. When any of these coils reaches a displacement that causes a reduction in the current sensitivity of the motor, higher distortion will result.

It has been empirically determined that, due to a magnetic fringe or leakage field at the pole tips, an excursion of 15% farther than the gap length results in a reasonable distortion level (approximately 3% harmonic distortion at low frequencies). The equal length voice coil, Fig. 2.24C, has the greatest potential for motor-generated distortion. However, it also yields the highest motor strength (the greatest total conductive mass in the highest density magnetic field). The equal length voice coil is a common configuration for compression drivers, where maximum excursion is intrinsically low. The underhung coil, Fig. 2.24B, allows greater excursion but requires a larger magnet due to the longer gap. For moderate flux density levels

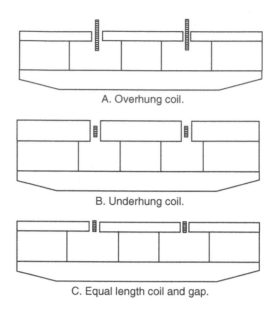

A. Overhung coil.

B. Underhung coil.

C. Equal length coil and gap.

FIGURE 2.24　Three basic voice coil/magnetic gap configurations.

(10,000 to 15,000 G), this design, as compared to the equal length design, requires approximately twice the magnet weight (twice the area and the same length) for a doubling of the gap length. This approximately doubles the excursion capacity, giving four times the acoustic power output capability (6 dB) for a doubling of magnetic weight (3 dB). The overhung coil, Fig. 2.24A, is capable of the greatest motor linearity, all else being equal. It is commonly seen on woofers used as direct radiators, where higher excursion is required. The major disadvantage here is that the coil that is not in the gap does not participate in transduction. The extra coil length does add both mass and dc resistance, however, reducing motor efficiency. In spite of this, there are numerous examples of successful commercial woofers using overhung coils. The transducer designer must take into account the often conflicting demands of high-efficiency, high-output, and low-frequency extension to arrive at an optimum design for a given range of applications.

The thermal limit of a magnetic loudspeaker motor is a function of the temperature limits of the materials used and heat transfer from the coil assembly to the outside world. Most adhesives used in the loudspeaker industry have an upper limit between 120°C and 177°C (250°F and 350°F). Some epoxy adhesives will tolerate higher temperatures, but they can require special curing processes and are therefore potentially more difficult to use. Wiring insulation may tolerate temperatures as high as 218°C (425°F). Anodized aluminum wire has the melting point of aluminum as a limit. Voice coils operated at high temperatures have higher resistance. A 1°C rise produces approximately a 0.4% rise in dc resistance in both copper and aluminum. Therefore, operating a voice coil 100°C above ambient (127°C or 261°F) will cause the voice coil resistance to increase to 40% above its ambient value. The following equation gives voice coil resistance at any temperature in degrees Celsius

$$R_T = R_o + 0.004(T - T_o) \qquad (2.1)$$

where,
R_T is the resistance at temperature T in ohms,
R_o is the resistance at ambient temperature T_o in ohms,
T and T_o have units of °C.

The operating temperature of the voice coil, T_{VC}, is determined by ambient temperature, the amount of power being dissipated in the coil, and a parameter called *thermal resistance*, expressed in degrees Celsius per watt, °C/W. The thermal resistance is a measure of the ability of an object to transfer heat away from itself. The lower the value of the thermal resistance, the more effective the object is at this transfer. As power is doubled, final temperature rise above ambient is doubled. Heat transfer in a loudspeaker is a function of the air gap design, voice coil design, and the ability of the loudspeaker frame and magnet to dissipate heat to the surrounding or ambient air. Referring to Fig. 2.25, the thermal rise T_{VC} of a stationary voice coil in an air gap is

$$\Delta T_{VC} = T_{VC} - T_s$$
$$= \frac{QL}{A_T K} \qquad (2.2)$$

where,
T_{VC} is the temperature of the voice coil in °C,
T_s is the temperature of structure (magnet) in °C,
Q is the electrical heating power (IR) in watts,
L is the effective air gap length in inches,
A_T is the total gap area in square inches exposed to the voice coil,
K is the conductivity of air or 7×10^{-4} W/°C.

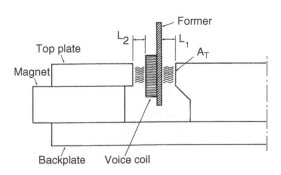

FIGURE 2.25   Heat conduction in magnetic loudspeakers.

As the air gap length is decreased and the area increased, heat transfer increases (or, equivalently, thermal resistance decreases). Making the voice coil former of aluminum will increase effective heat transfer area; the thicker the aluminum, the greater the effect. Voice coils wound on aluminum formers with large diameters in magnets with large gap areas and very tight coil to gap tolerances are capable of handling high electrical power due to good heat transfer in the air gap. In short, large, accurately constructed loudspeakers can usually handle more power. As the loudspeaker moves, it may be able to pump the air in the gap to improve heat. The loudspeaker designer may be able to exploit this behavior. Given voice coils of the same length, the underhung and equal-length configurations will have greater heat transfer capacity. The overhung coil would only conduct heat well in the gap region, while the coil ends remaining out of the gap would be more likely to suffer damage at high power level because of relatively poor heat transfer. Typical thermal behavior for most coils is on the order of 0.5°C/W to 3°C/W input.

A heat-conducting magnetic liquid may be used to improve heat transfer. Known as *ferrofluids*, these fluids will be retained in a magnetic air gap due to magnetic attraction. Their thermal conductivity is seven to ten times higher than that of air. Since ferrofluid alters the mechanical damping of the moving assembly, its use has implications for the design of the motor assembly. There are also issues related to compatibility of ferrofluid with adhesives and materials used in the construction of a transducer. For these reasons, ferrofluids should generally be designed into a loudspeaker, rather than added on.

Temperature rise in voice coils is not instantaneous. It is directly related to mass. As one might suspect, light voice coils have short thermal rise times, and vice versa. The thermal time constant of a loudspeaker coil (the time required for the coil to reach 63% of its final value) is given by:

$$t = MC\frac{\Delta T}{Q}$$
$$= MC\frac{L}{A_T K} \tag{2.3}$$

where,

t is the time constant in seconds,

M is the mass of the coil,

C is the specific heat of voice coil material in joules per
    gram in degrees Celsius,

$9T/Q$ is the thermal resistance in degrees Celsius per watt.

For example, a typical copper woofer voice coil has a mass of 24 g and a gap heat transfer coefficient (thermal resistance) of 1°C/W. Copper has a specific heat of 0.092 cal/g°C or 0.0220 J/g°C. Therefore, using Eq. 2.3, $t = 0.528$ s. This is a typical voice coil response time. An aluminum coil will typically have a shorter thermal time constant.

The time constant of the magnetic structure and frame can be on the order of hours. For this reason, long duration power tests are required to evaluate the maximum power tolerance of transducers. Initially, the voice coil might be at 280°F (137°C), but over the course of 2 hours, the mechanical structure (typical 1 to 3°C/W) could rise another 200°F to 300°F (100°C to 150°C), bringing the voice coil well over the thermal limit of its materials and adhesives. Heat transfer from the frame and magnet to the air is another important consideration. Although the rise time is large, the final temperature may vary greatly due to the enclosure. A vented enclosure with vents at the top and bottom with no fiberglass insulation might provide adequate ventilation for a hot loudspeaker. The same loudspeaker in a closed box stuffed with fiberglass might be subject to a dangerously high temperature rise. Attention to this final thermal path is warranted in applications that will demand maximum output from enclosed loudspeakers.

The efficiency of a loudspeaker has a direct bearing on the thermal load it must withstand for a given acoustic output level. The more efficient the loudspeaker, the lower the self-heating for a given output level; all else being equal, a loudspeaker with 3 dB higher overall sensitivity for a given impedance will experience one-half the thermal load for a desired output level.

In concert touring use, loudspeakers are routinely operated at and even beyond their design limits. Given that a loudspeaker that operates at twice its voice coil resistance due to heating will be 6 dB less sensitive, sound quality can vary greatly over the course of a performance. In failure situations, the nature of the input signal will usually determine the type of failure mode. Thermal failure can be precipitated by compressed high-frequency content material (low dynamic range). Mechanical failure is often due to dynamic, percussive material, such as might occur in a recording studio with drum channels set to solo, as well as other signals that do not limit dynamic range. Another cause of mechanical failure, most often in high-frequency transducers, is the application of a highly clipped signal that has been passed through a high-pass filter. Such a signal will contain a peak-to-peak voltage that is twice that of the input signal. This phenomenon is illustrated in the section on crossovers.

2.6.2 Heat Transfer Designs for High-Power Woofers

Of all the components in a sound reinforcement system, more heat is generated in low-frequency devices than in any other. While high-frequency horn driver combinations deliver 110–117 dB/1 W/1 m and midrange devices deliver 100–110 dB, woofers rarely exceed 100 dB. A typical woofer in a vented enclosure is in the 94–97 dB/1 W/1 m range. These devices are typically 2–8% efficient. The remaining 92–93% of the power goes directly into producing heat. Adding to the problem is the fact that much modern program material is bass-heavy.

As understanding of heat transfer mechanisms in loudspeakers grew, designs appeared that improved heat transfer from the voice coil and gave improved thermal power handling ratings, Fig. 2.26.

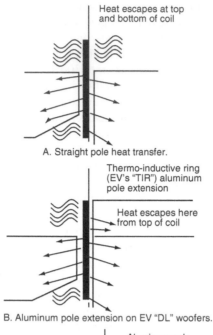

A. Straight pole heat transfer.

B. Aluminum pole extension on EV "DL" woofers.

C. Improved full-coil air conduction in EV "EVX" woofers.

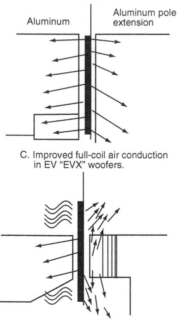

D. JBL "vented gap" natural forced-convection heat transfer.

FIGURE 2.26 Various coil/gap geometries showing the evolution of heat transfer designs in modern woofers.

The heat transfer methods discussed here are simply methods to transfer heat away from the voice coil. If there were no heat transfer paths out of the magnetic circuit, the speaker's temperature would continue to rise without limit. In the cases of drivers on exposed horns, natural convection transfers sufficient thermal energy to prevent overheating. The thermal resistance of the direct convection transfer path is on the order of 1–2°C/W. In the case of a woofer in a fiberglass-lined enclosure, this resistance may be five times greater. Heat buildup can be substantial. This mechanism is often ignored. Several proprietary loudspeaker systems have been developed in an attempt to address this problem, but most sound reinforcement systems still provide no designed-in mechanism for transferring heat out of the enclosures.

2.7 RADIATOR TYPES

In addition to converting electrical energy to mechanical energy, a loudspeaker must include a means for converting mechanical energy (e.g., the motion of a diaphragm) into acoustic energy. For purposes of this chapter, the various means for accomplishing this final conversion are referred to as *radiators*. While there is overlap in our definition of the terms *transducer* and *radiator*, it is extremely useful in understanding loudspeakers to consider the function of acoustic radiation as a separate subject from electroacoustic conversion. In general, there are two broad types of radiators: direct radiators and horn radiators.

2.7.1 Direct Radiators

The simplest form of radiator is the direct radiator, in which the diaphragm is directly coupled to the air. Most hi-fi loudspeakers consist of combinations of different sizes of direct radiators. Various forms of direct radiators were described in the previous section. In this section, we will outline their acoustic attributes.

2.7.2 Cone Radiators

In a cone radiator, the diaphragm is in the shape of a truncated cone. The concave surface of the cone is usually, but not always, the one which radiates sound. The cone shape is partially dictated by expediency: it allows the magnet structure to reside at the rear of the transducer assembly, while at the same time allowing for the use of a spider and a surround to suspend the diaphragm. This dual-element suspension provides positive centering of the voice coil in the magnet gap, and it helps constrain the motion of the cone to the desired linear path.

The above notwithstanding, there are also acoustic motivations that tend to favor the cone shape. At first glance, one might expect a flat piston to offer superior on-axis response and directivity to a cone. A cone shape is generally preferable, however, when the excitation will be applied near the center of the diaphragm. Due to the fact that sound propagates at a finite velocity in a solid, the motion of the outer portion of the diaphragm will follow the initial excitation by some amount of time. If the diaphragm were flat, radiation from the outer portions of its surface would arrive at an on-axis observation point at later times than radiation from the center. The cone shape reduces the distance that must be traveled by sound radiated from the outer portions to on-axis listening positions. Since the velocity of sound in the cone material is typically greater than the velocity of sound in air, a cone shape having the optimum included angle will tend to synchronize on-axis radiation from the outer portions of the diaphragm with that from near the center. Given a judicious choice of angle, the useful range of response of a cone transducer can be extended to a significantly higher frequency than would otherwise have been the case.

2.7.3 Dome Radiators

Another variant on the direct radiator theme is the dome radiator. Most often, this radiator takes the form of a convex dome driven and suspended at its periphery. The material used to form the dome may be soft, as is the case with coated synthetic textiles, or hard, in the case of metals or composite materials (e.g., carbon fiber/epoxy). Dome radiators are most popular for high-frequency elements, although a number of dome-shaped midrange elements are also available. As with the cone radiator, the convex shape of the typical dome radiator has acoustic motivations. Since the excitation is at the edge, the dome's mechanical motion will propagate inward. As a result, if the shape were flat, radiation from the inner portion would arrive at an on-axis observation point later than radiation from the edge. The convex shape helps deliver a more coherent wave front to an on-axis listening position. It is common practice to suspend a small round cover just in front of the center of the dome.

2.7.4 Ring Radiators

Yet another form of direct radiator is the ring radiator. In a typical ring radiator, a flexible ring-shaped diaphragm is rigidly captured along its inner and outer circumferences and driven along a concentric circular line between those two circles. There is, then, no distinction between the diaphragm and the suspension, as a single part fills both functions. A dome tweeter with a cover over the center of the dome functions as a ring radiator. Ring radiators can also be used to drive short horns.

The JBL 075 Bullet is an example of a ring radiator. Intended for use above 7 kHz, the diaphragm is a V-shaped ring of aluminum attached to a voice coil and former.

Figure 2.27 shows a ceramic version of the ring, which is made by Yamaha. The phase plug is a simple slit ending in a large enough mouth to project the desired low end of the driver. The suspension is the diaphragm itself, and it is quite stiff. Ring radiators are typically operated in or above the principal resonance frequency of the diaphragm assembly.

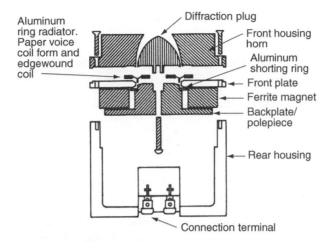

FIGURE 2.27 Yamaha ceramic-magnet ring radiator cross section. Courtesy Yamaha International Corp.

2.7.5 Panel Radiators

Both electrostatic and *planar* electrodynamic speakers fall into this category. As with the ring radiator, there is a mixing of functionality between the diaphragm and the suspension. The acoustic advantage, at least in principle, of a panel radiator, is that the driving force is applied uniformly over a large portion of the diaphragm. For this reason, diaphragm rigidity is not an essential design element, as is the case with cone radiators. An interesting characteristic of a large panel radiator is that it will essentially project a shadow of its shape as a listening pattern; this shadow of the speaker's radiation pattern will take up a large part of a typical listening area, particularly at close listening distances. This is claimed to produce a wider "sweet spot" compared to conventional cone systems.

2.7.6 Horns

Horns are used to increase the efficiency of a transducer and to control the directivity of the sound that is radiated. Horns are characterized by a number of parameters. The earliest approach to a predictive model, and the one still employed in acoustics texts, is characterization by the rate of increase of cross-sectional area with longitudinal position in the horn. Other means of characterization are related to the shapes formed by the horn walls.

Of all possible expansion (or flare) rates, a relative few have found use in horn design and analysis. Those most commonly encountered are exponential, hyperbolic, conic, and catenary. In general, the change of cross-sectional area with position in a horn can be expressed as

$$A(x) = F(x) \tag{2.4}$$

where,

$A(x)$ is the cross-sectional area at a point x along the axis of the horn,
$F(x)$ is some function of x.

For example, in an exponential horn,

$$A(x) = A_0 e^{mx} \tag{2.5}$$

where,

A_0 is the area of the horn at its throat or entry,
m is a constant called the *flare rate*.

Much more detailed information is available on the subject of the acoustic characteristics resulting from different rates of expansion from the sources cited in the Bibliography. The models are useful in analyzing the propagation of acoustic energy within a horn, but other considerations become dominant in determining the nature of the radiated sound beyond a horn's mouth.

For this reason, practical horns have come to be known more by salient details of their sidewall shapes than by their flare rates. The more common types are described below.

2.7.6.1 Radial Horns

Radial (or sectoral) horns were claimed to allow a natural radial expansion of the sound wave from the driver, while maintaining an exponential expansion rate. Typically, a radial horn has straight horizontal sides and top and bottom walls that are in the form of spherical *sectors*. The design approach employed for a radial horn involves positioning the sides at approximately the desired angle for horizontal coverage. Given the area expansion desired, the top and bottom surfaces are then derived mathematically. The most popular materials used in making radial horns are cast aluminum (now relatively uncommon), molded plastic, laminated glass fiber, and polyester resin. This type of horn was in widespread use from the 1930s until approximately the mid-1980s, by which time constant directivity types had become more popular.

Figure 2.28 is an Altec 311-60; it has a 60° horizontal coverage and is intended for use above 300 Hz, using a 1.4 inch driver. Altec was well known for this design, with its characteristic vertical vanes at the mouth of the horn.

2.7.6.2 Multicell Horns

Multicell horns were the first horns to be employed specifically for their directivity control attributes. The design approach was straightforward—several small horns were affixed together in an array, with each horn to supply a portion of the total coverage angle. These small horns were connected to a common

FIGURE 2.28   Altec Lansing 311-60 cast aluminum sectoral horn with sound-deadening material. Courtesy Altec Lansing Corp.

FIGURE 2.29   Altec Lansing 1.4-inch throat, all soldered and coated steel horn family showing throat plumbing fixtures. Courtesy Altec Lansing Corp.

manifold so that a single driver could power them, Fig. 2.29. Multicell horns first came into use in the late 1930s. They were originally made of sheet metal soldered together and either filled on the outside with sand or covered with a mechanical damping material.

2.7.6.3 Controlled Directivity Horns

The first constant directivity type of horn appeared in 1975. Developed by Electro-Voice, they employed a hyperbolic-flare throat section coupled to a conical radial bell section, as shown in Fig. 2.30. This horn shape yielded good low-frequency loading and relatively constant angular beam width in both vertical and horizontal directions over a wide frequency range. At the time, its design represented a major departure from previous thinking. Don Keele, the designer

FIGURE 2.30 Electro-Voice HR9040 constant directivity horn. Courtesy Electro-Voice, Inc.

of the horns, presented an AES paper ("What's So Sacred about Exponential Horns"). In the paper, he disclosed several empirically developed relationships between mouth size, frequency, and maintenance of coverage angle. The concept of a waveguide as applied to an acoustic radiator was used as the basis for predicting and controlling the directivity of a horn.

Keele's paper and horn designs provided impetus for further empirical investigations of controlled directivity horns. Altec Lansing, at that time a competitor of Electro-Voice, introduced a family of horns using a narrow, vertical diffraction slot located at an intermediate point in the horn. With the appropriate choice of location for this slot, it is possible to make a horn with any desired combination of sidewall angles and aspect ratio (relationship between the height and width of the mouth).

This family of devices was dubbed "Manta-Ray," and a number of designs based on this thinking were introduced over the ensuing years, Fig. 2.31.

Another approach to achieving the goal of frequency-independent directivity was represented in the JBL *biradial* family of horn designs. Also developed by Don Keele, who had by then taken an engineering position with JBL, the biradial shape employs continuously varying flanges in both directions, ending in a continuous horn. The vertical diffraction slot was retained. An exponential expansion rate was part of the design, and vertical and horizontal radial bell shapes (thus the term *biradial*) were employed. Three biradial horns are shown in Fig. 2.32. They are fabricated from cast aluminum (throat section) and molded fiberglass (bell section) and fit 2 inch exit drivers.

2.7.6.4 Voice Warning Horns

Figure 2.33 shows another variety of controlled directivity horn, designed by Bruce Howze of Community Professional Loudspeakers, and originally built for Whelen Engineering. The horn used a slightly different directivity control philosophy: a controlled horizontal pattern (45° and a narrow vertical pattern), due to the 70 in (1.78 m) vertical dimension. The horn uses 16 siren drivers and

FIGURE 2.31   Altec Lansing Manta-Ray horn family, cast aluminum throat and soldered, coated bell construction. Courtesy Altec Lansing Corp.

FIGURE 2.32   JBL biradial horn family cast aluminum throat and fiberglass bell construction. Courtesy JBL/UREI.

the system, with a 1600 watt input, generates 127 dB at 100 m (328 ft). It is used in a similar manner to a searchlight: aim and shoot via a remote rotor.

2.7.6.5 Asymmetric Directivity Horns

As the ability to determine the optimum loudspeaker directivity requirements for specific applications was refined, it became apparent that the required directivity was usually not symmetrical about a horizontal plane through the axis

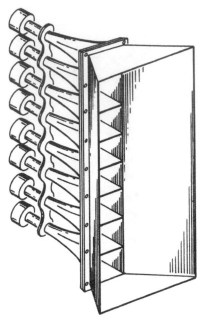

FIGURE 2.33   Whelen Engineering horizontal diffraction horn with multiple drivers. Courtesy Whelen Engineering.

of the horn. From the point of view of the loudspeaker, it is most common for the required horizontal coverage to be relatively narrow at the greatest distance from the source and to become successively wider at closer distances. In addition, it is desirable for the vertical angle of greatest intensity to be as large as possible—i.e., for greater energy to be directed to the seats at the greatest distance from the loudspeaker in order to produce similar SPL values throughout the audience. One early attempt to address this requirement was the JBL 4660, shown in Fig. 2.34.

Another design, developed for a specific application, is the IMAX® PPS (Proportional Point Source) loudspeaker, developed by the author. Figure 2.35 is the high-frequency horn used in this loudspeaker, and Fig. 2.36 is its 4 kHz isobar.

Dave Gunness, chief engineer at Electro-Voice at the time, developed a family of asymmetric directivity horns in the late 1980s and early 1990s. These were known as *Vari-Intense* devices.

Optimized (asymmetric) directivity is an attractive engineering goal, but there are a number of obstacles to its widespread acceptance:

1. Computer-based sound system prediction software is required in order to visualize its effectiveness and optimize aiming and device placement.

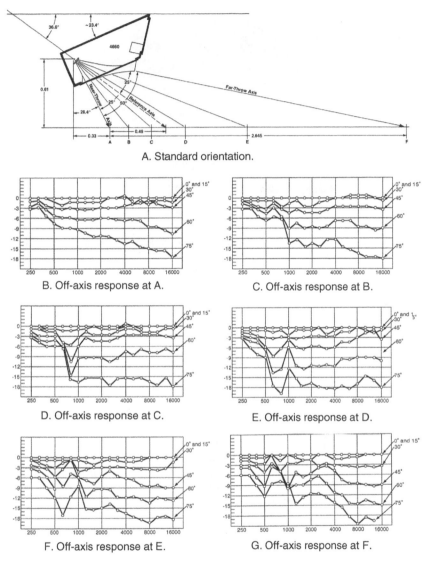

FIGURE 2.34 JBL 4660 asymmetric-directivity horn. Courtesy JBL.

2. Exactly what constitutes ideal directivity is a strong function of the space in which the device is to be used. The ideal directivity will vary, for example, for different loudspeaker elevations within the same space.

3. With the exception of the proprietary IMAX® loudspeaker, there are currently only high-frequency devices available with this type of directivity. Achieving uniform sound pressure levels throughout the seating, but at high frequencies only, is of limited value.

FIGURE 2.35 IMAX® Proportional Point Source high frequency horn. Courtesy James E. Mitchell & Associates.

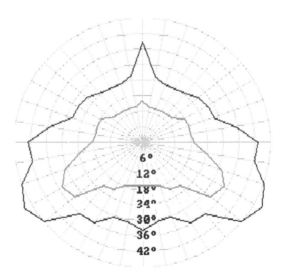

FIGURE 2.36 4kHz isobar of the loudspeaker in Fig. 2.35. Courtesy James E. Mitchell & Associates.

2.7.6.6 Acoustic Lenses

Although an acoustic lens is not generally regarded as a directivity control device, it can function as a directivity alteration device. While acoustic lenses are used to widen a pattern, they can also be used to narrow a horn's directivity. An acoustic lens is usually formed with parallel plates of strategically chosen shapes placed at an angle to the direction of sound propagation. Differing path lengths through different portions of the lens create arrival time relationships for the associated components of the wave that generate specific directivity characteristics.

The slant-plate lens assembly, shown mounted on a JBL studio monitor in Fig. 2.37, is one notable implementation of an acoustic lens. Note that the device has concave openings in the plate array. As the wave leaves the horn and progresses through the lens plate array, the center of the wave reaches the air on the outside first, due to the shorter path through the lens. The outer portions of the wave travel through longer paths within the lens and are therefore delayed in time relative to the portions that came from the center. The net effect is to produce arrivals that are better synchronized—and therefore stronger—at positions that are off axis in the horizontal, thus widening the polar pattern in that direction. The vertical pattern would ideally be unaltered. Lenses have the undesirable property of causing relatively strong reflections back into the horn.

FIGURE 2.37 JBL studio monitor employing slant-plate-type acoustic lens on a high-frequency component. Courtesy JBL.

2.7.6.7 Folded Horns

One of the practical drawbacks of horns, particularly those intended for use at low frequencies, is their physical size. Folded horns were developed in response to this problem and have been in use in various forms for more than half a century. A folded horn is produced by truncating the shape at point, providing a reflecting surface to change the direction of the outgoing wave, and continuing the horn's expansion in another direction, usually opposite the prior one. Successive horn sections are typically positioned outside their predecessors. As many reversals are generated as are necessary to create the desired path length and mouth size. There are, as with other horn types, many variations on folded horns.

Figure 2.38 shows a University Sound GH directional trumpet cross section and how the area expands by making two 180° turns. The design was introduced in the 1940s. Another University folded public address horn design is the cast zinc Cobraflex, shown disassembled in Fig. 2.39. This horn expands into a double mouth.

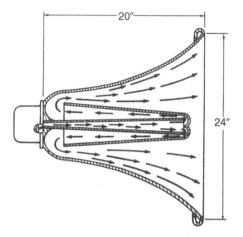

FIGURE 2.38   Cross-sectional drawing of an exponential folded horn.

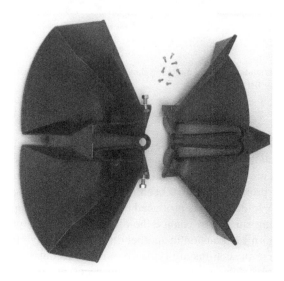

FIGURE 2.39   University Cobraflex horn disassembled. Courtesy Altec Lansing Corp.

One of the most recognizable low-frequency horns is the Klipschorn, shown in Fig. 2.40. It was named for its inventor, Paul Klipsch, who was one of the pioneers in horn loudspeaker design. The Klipschorn uses a single 15 inch loudspeaker in a relatively compact package. It is designed for placement in a corner of the room, with the room's walls forming an extension of the horn shape.

The Cerwin-Vega E horn is another form of folded bass horn. An 18 inch low-frequency driver sits between the upper and lower mouths and faces to the rear in a compression chamber. The E horn makes one 180° fold that opens to the double mouths. The horn is intended for use with additional mouth extensions and in multiples for low-frequency coupling.

The W horn is another folded bass horn design. The best-known example is the RCA theater horn, which uses two 15 inch drivers. The W uses forward-facing drivers, and the horn flare is designed as a W-shaped double fold, expanding to twin mouths.

The Altec 31A uses a single 90° fold. This allows for a short front-to-back dimension. Additionally, the driver is mounted facing downward, making it rain and dust resistant. This horn uses a 120° mouth and is used above 500 Hz for a wide variety of applications, including voice-only systems.

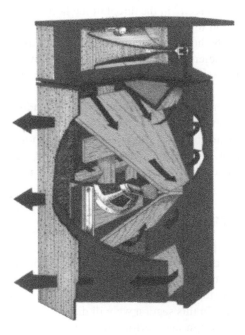

FIGURE 2.40 Klipschorn folded corner bass horn—rear cutaway view with high-frequency components. Courtesy Klipsch & Associates.

The obvious advantage of a folded horn is the reduced package size for a given horn length. This advantage is offset by the fact that, for each reversal fold in the horn's shape, a reflection is generated inward, opposite the desired direction of wave propagation. These reverse waves are reflected again in a forward (outgoing) direction when they reach the horn's driver area, generating late signal arrivals that cause significant deviations from ideal in the horn's response. For this reason, folded horns generally find use in applications that are relatively undemanding of fidelity.

2.7.6.8 Special Considerations for Low-Frequency Horns

Over the years, a number of horns have been developed specifically to radiate low frequencies. In the past, the primary motivation for the use of a horn to reproduce low frequencies was improved efficiency as compared to a direct radiator. The current availability of power amplifiers with extremely high output capacities and woofers that are capable of utilizing that power has rendered the issue of efficiency less important than that of size. As a result, there are fewer low-frequency horns on the market today than in the past.

The most common difference between bass horns and those intended for mid- and high-frequency use, aside from the bass horns' larger size, is that low-frequency horns typically do not employ compression drivers. Instead, a cone transducer is mounted directly in the throat of the horn.

A potential issue in low-frequency horn design and operation is the transitional behavior of the horn. It is common practice to use a bass horn/driver combination to a sufficiently low frequency that the horn is too small to provide substantial acoustic loading in the lower portion of the bandwidth of use. In this frequency range, the driver must operate as a direct radiator, with correspondingly lower sensitivity. Although it has been asserted that this discrepancy, which can exceed 10 dB, may be overcome through the use of ports, in actuality the only means of leveling the device's response between the two regimes of operation is with equalization of the input signal. Given proper equalization, a bass horn may be used in this fashion with excellent results.

2.8 LOUDSPEAKER SYSTEMS

Most practical loudspeakers are systems comprising multiple transducer/radiator subsystems, each of which radiates a portion of the audio-frequency spectrum. This area of loudspeaker design has a major impact on a loudspeaker's ultimate performance, yet this portion of the design process is frequently shortchanged. In this section we will discuss some considerations for loudspeaker system design and performance and provides some illustrative examples.

The desirability of dividing the audible frequency range into multiple bands is taken for granted in most loudspeaker applications. The most compelling reasons for dividing the spectrum among multiple components are:

1. By itself, the bandwidth of a practical transducer/radiator is inadequate to meet the bandwidth requirements for a complete loudspeaker.
2. The directivity of a single transducer/radiator will not be sufficiently consistent with frequency to meet reasonable goals for the directivity of a full range loudspeaker.
3. The maximum available acoustic output of a single transducer is inadequate. Sharing the output demand among a number of band-specific components enables a loudspeaker to produce greater total acoustic power.

In designing a loudspeaker system, one should, at the very least, have a working knowledge of the disciplines involved in the design of the component parts. The system designer's challenge is to make a collection of individual components function as a cohesive whole while meeting the cost, size, and aesthetic requirements of the loudspeaker's intended applications. The design of a successful loudspeaker system involves much more than simply selecting a group of components and building a box to house them.

It is axiomatic that, in addition to the required technical expertise, a loudspeaker designer should have the capability of subjectively evaluating a loudspeaker's performance—critical listening—and that the final determinant of a loudspeaker's success will almost always be subjective acceptance. It is equally true that there are always objectively observable phenomena that correlate with subjective preferences. The difficulty in reconciling the two is a direct result of the very large number of objective elements that must be accounted for in order to fully characterize the performance of a loudspeaker. This subject is covered in greater depth in the "Characterization of Loudspeaker Performance" (Section 2.9).

Loudspeaker systems are often categorized by the number of spectral divisions made in the system, as in two-way or three-way systems. Generally speaking, a loudspeaker system consists of two or more transducer/radiator combinations, a crossover network, and an enclosure that houses everything. In addition to providing a convenient package for the components, the enclosure serves structural, acoustic, and aesthetic purposes. The sections on acoustic boundaries and electroacoustic models provide information about some of the acoustic effects of enclosure design.

2.8.1 Configuration Choices

A number of decisions about a loudspeaker's configuration are typically made early in the design process. These include:

1. The number of spectral bands, or divisions.
2. The type of radiator to be used for each band.
3. The location and orientation of the individual components within the system housing.

In determining the number of frequency bands to be used in a loudspeaker, several conflicting demands must be reconciled. Choosing a greater number of divisions creates the possibility of greater broadband acoustic output and more optimal radiator configurations for each band. On the other hand, each added band adds to the size, complexity, cost, and more often than not, to nonideal aspects of the acoustic behavior of the finished design.

The type of radiator chosen for each band is often a matter of custom or convention rather than of engineering. Where possible, it is generally desirable to match efficiencies and directivities of adjacent bands over a range of frequencies centered about their crossover point. This is most readily accomplished when similar types of radiators are used for both of the bands in question.

The location and orientation of individual components is an area worthy of careful attention. It is common practice to place all of the transducers on a flat panel (a baffle), displaced from each other in vertical and/or horizontal directions. In the case of loudspeakers designed for stereo reproduction, it is also common practice to make pairs of speakers in a mirror-image layout.

The aforementioned common practices have developed over many years, with the primary motivation being cost and ease of manufacture. Another approach to loudspeaker system design is the coaxial layout. First employed in the earlier part of the 20th century, this practice involves locating two or more bands of a loudspeaker along a common axis. While the coaxial approach is typically more difficult to implement, it has some advantages over more conventional layouts.

2.8.2 Types of Loudspeaker Systems

The simplest form of loudspeaker employs a single full range transducer to reproduce all frequencies. The most common applications for this type of device are limited-bandwidth (e.g., speech) systems and inexpensive music reproduction systems.

For residential music systems, one of the more common configurations is a two-way system utilizing a small (typically 6 or 8 inch) woofer and a dome tweeter. Figure 2.41 is one such system. More elaborate (and costly) systems are also employed for residential use, some employing line arrays (see Section 2.8.4) of transducers for one or more of their bands. As is the case with professional loudspeakers, visual aesthetics can play as important a role as performance in setting design requirements.

One of the more common types of loudspeakers for general sound reinforcement use is a two-way system consisting of a direct radiator woofer and a horn/compression driver high-frequency subsystem, with the components being located one above the other on the front face of the enclosure. The typical package for this type of loudspeaker is a trapezoidal enclosure. The trapezoid designation describes the plan view of the enclosure, and the shape

FIGURE 2.41   Two-way monitor speaker. Courtesy Genelec.

allows multiple loudspeakers to be arrayed in the shape of an arc segment, with the included angle between adjacent loudspeakers being equal to twice the sidewall angle. A large number of manufacturers offer loudspeakers that fit this description. Typical sidewall angles range from 12° to 15°, while the horizontal coverage angle (the angle at which the output has fallen to 6 dB below the level on axis) of the high-frequency horns used in such devices is typically either 60° or 90°, and the coverage of the woofer is entirely uncontrolled.

2.8.3 Performance Issues in Multiway Systems

Before trying one's hand at designing a multiway loudspeaker, it is a good idea to develop a familiarity with the complete audio signal chain and to understand the implications each decision will have on the acoustic signal that will reach the listener's ear. Figure 2.42 is a functional diagram of the electrical and acoustic signal paths from the source (electrical input signal) to the observation (listening) point.

In the above representation,

$$F_T(S, x, y, z) = \sum_N \left[F_n(S) G_n(S, x, y, z) \right] \tag{2.6}$$

where,
F_T is the total electroacoustic transfer function,
F_n is the (electrical) transfer function of the nth crossover filter,
S is the Laplace complex frequency variable,
G_n is electroacoustic transfer function of the nth radiator in the system,
x, y, and z are Cartesian spatial coordinates.

The concept is general and will accommodate an arbitrary number of spectral divisions.

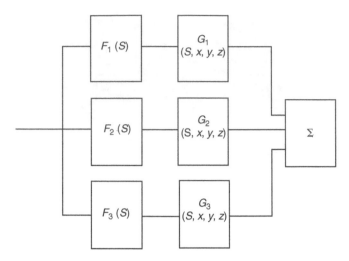

FIGURE 2.42   Functional diagram of multiway loudspeaker.

It should be noted that, although this diagram depicts a loudspeaker with a passive crossover, it might also be used to represent an active system simply by including gain in the transfer functions of the crossover blocks. An active loudspeaker simplifies the task of crossover design. Since the power amplifier serves as a buffer between the crossover and the transducers, the frequency-dependent impedance behavior of the transducers becomes a very small factor in the design process. Note that the crossover filters are in cascade with the transducers, while the acoustic outputs of the devices are summed acoustically at the listening position. The nature of such a multipath system is complex, and detailed prediction of its response at every likely listening position is a nontrivial task.

Note also that the transfer functions G_n are functions of the spatial coordinates x, y, and z as well as of the complex frequency variable S. This spatial dependency includes the effects of source directivity, propagation delay, and the inverse square law. If the system is not coaxial (and sometimes even if it is), then the lengths of the paths from each transducer to a given listening position will not generally be the same. The most common practice is to choose an axis along which one will attempt to equalize these acoustic path lengths and then to optimize the speaker's behavior on this axis. In the case of a two-way loudspeaker with the transducers displaced along a line in the plane of the baffle, it is possible to make path lengths equal at every point in a plane. Once more than two spectral divisions are present, even this limited goal is no longer possible. It may well be the case that, in a three- or four-way system, there is no point for which all acoustic path lengths from the transducers to a listener's ears will be equal.

The effect of unequal acoustic path lengths is that signals from different radiators will reach the listener at different times, even though they originated simultaneously. This timing discrepancy is not generally sufficient to be recognized by a listener as comprising distinct multiple events, but it is enough to have audible and undesirable effects on a loudspeaker's amplitude response, as well as on its ability to reproduce transient signals. Even when an axis or plane exists in which signal synchronization has been achieved, positions off the axis or outside of the plane will not receive the benefits of such synchronization.

The subject of complex addition of time-varying signals is beyond the scope of this chapter, but it is dealt with in many introductory circuit analysis texts. Additional effects of signal synchronization, and the lack thereof, on loudspeaker response are illustrated in the section on crossovers.

There are a number of ways in which the problems caused by noncoincident transducer locations may be addressed by a loudspeaker designer. One common approach is simply to assert that the response anomalies caused by this configuration are not audibly significant and to accept (or avoid acknowledging) their presence. Another is to employ crossover filters with very steep slopes so as to minimize the frequency range over which anomalies due to path length differences will be present. As will be shown in the crossover section, the latter technique has its own set of drawbacks and may in some cases create more serious problems than it solves.

One means of addressing the synchronization issue is with a coaxial loudspeaker. This type of loudspeaker is most often two-way, although it is possible to design a three- or four-way coaxial system. There are benefits in making the midrange and high-frequency components of a three-way system coaxial, while leaving the low-frequency portion displaced in the more conventional manner.

A coaxial loudspeaker will always possess symmetrical response behavior, Fig. 2.43. That is, the response at a given angle from its axis will be mirrored at the same angle in the opposite direction. Additionally, if the acoustic path lengths from transducer to listener are equal on the system's axis, it is possible to preserve this synchronization at all listening positions with a coaxial design. Even though it is possible to achieve signal synchronization over a wide angular range with a coaxial loudspeaker, this possibility is not always realized in practice. When a coaxial loudspeaker fails to achieve coincident performance (i.e., it fails to behave as a single full range radiator), its sole distinction as compared to more conventional configurations is that frequency-dependent anomalies related to crossover interactions will be symmetrically located about the loudspeaker's axis.

Figure 2.44 illustrates some of the effects caused by displaced transducers. The frequency selected for display is the closest ⅓-octave band center to the crossover frequency of the speaker. Figure 2.45 shows the improved polar behavior that can be produced with a coaxial loudspeaker, again with the ⅓-octave band center chosen so as to most closely match the crossover frequency.

FIGURE 2.43   Small two-way coaxial loudspeaker. Courtesy Frazier Loudspeakers.

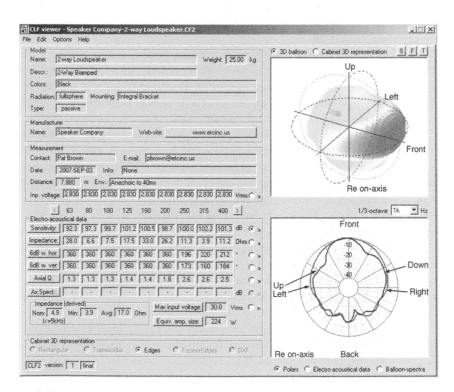

FIGURE 2.44   Generic two-way loudspeaker directivity balloon and polar pattern in the crossover region.

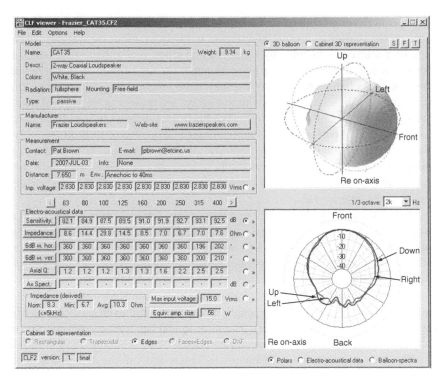

FIGURE 2.45   Coaxial two-way loudspeaker directivity balloon and polar pattern in the crossover region.

Note that the polar pattern of this speaker is still not perfectly symmetrical, even though the transducers are coaxial. This is due to asymmetric placement of the coaxial woofer/tweeter assembly within the enclosure. The effects of enclosure design on loudspeaker performance will be covered in more detail in the section on acoustic boundaries.

2.8.4 Line Arrays

Another type of loudspeaker system is a line array. Although line arrays have much in common with other types of loudspeaker systems, they have some attributes that are unique enough to justify their separate treatment. A line array may form a complete full-range loudspeaker or one or more bands thereof. In a line array, individual radiators are arranged in a straight line or an arc segment. It is also possible for a number of complete loudspeaker systems to be configured as a line array. It is this configuration that has come into fashion in recent years. In the simplest form of line array, each of the elements—usually a small cone transducer—is supplied an identical full-range signal. This type of array, also called a *sound column*, was popular in this country through much of the 1970s and is still in common use in installed sound systems.

Recent developments in DSP technology, combined with the constant pressure on the touring concert reinforcement industry to minimize weight, blockage of audience sight lines by speakers, and truck space, have resulted in a resurgence of interest in line arrays. As attractive as some of their perceived performance characteristics may be, they have inherent limitations. First, the directivity attributes associated with line arrays are present in the vertical plane (along the length of the array) only. The horizontal directivity is only as good as the horizontal performance of the individual devices used to form the array. Secondly, line arrays invariably comprise discrete elements, as opposed to a continuous line source. This periodicity exacerbates problems with nulls and lobes, and it causes the off-axis impulse response of a line array to contain multiple discrete arrivals.

It is often incorrectly asserted that a line array behaves, or can behave, as a line source. A line source is largely a theoretical construct. It consists of a long, narrow radiator that radiates sound with perfect uniformity at every point on its surface. This assumption of perfect uniformity, while impossible to achieve in practice, simplifies the mathematics required to model the behavior of a line source. When used for illustrative purposes in texts, line sources may additionally be assumed to have infinite length, making possible even further simplification of the mathematical model. The same model has been employed in texts on electromagnetic theory, for the same reasons.

The two assumptions—continuous radiation and infinite length—lead to two interesting results. First, due to symmetry, the frequency response of an infinitely long, continuous line source is not a function of observation position along the line. For example, if the line is assumed to be coincident with the Z-axis in a cylindrical polar coordinate system, then its response will not vary with changes in the Z-coordinate of an observation position (i.e., for movement in a direction that is parallel to the line). Second, due to the infinite length of the source, the wave front (a collection of isophase points) will form a cylindrical, rather than a spherical, shape. For this reason, the intensity of radiation in the outward direction falls off as the inverse of the first, rather than the second, power of the distance from the line.

As interesting and attractive as the two above results may be, they are not achievable in any physically realizable array. The effects of radiation that is neither continuous nor uniform, and of finite array length, cannot be neglected in discussing the behavior of real-world systems. Unfortunately, these issues have been glossed over or completely ignored in the information that is provided regarding the performance of commercially available line array products.

Full-range line arrays characteristically have relatively narrow vertical radiation patterns. The details of these radiation patterns vary widely with frequency and typically contain undesirable off-axis nulls (deep response notches) and lobes (response peaks). The same phenomena that produce off-axis response variations in a noncoaxial, multiway loudspeaker—interference caused by variations in the relative distances between multiple sources and the listener—create this directivity. At high frequencies, the angular separation between the first two nulls—and therefore the useful coverage angle—may be on the order of 5° or less.

A number of remedies to the problem associated with line arrays have been implemented over the past 50 years. There are two primary areas in which the line array intrinsically poses challenges to the designer: total array length and individual device spacing. Both must be addressed in order to produce a well-behaved system.

One means to address the issue of total array length is to implement a tapered array. In this type of array, only the innermost elements carry the highest frequencies. The signals applied to the more outwardly placed elements in the array are low-pass filtered at successively lower frequencies. The goal of this approach is to make the effective length of the line array become shorter at higher frequencies. An alternative way of stating this goal is that one desires the ratio between the effective length of the array and the wavelength of sound to be invariant. With the ability via DSP processing to create filters of essentially arbitrary amplitude and phase response, it has become relatively straightforward to create tapered arrays. Additionally, the availability of frequency-independent delay makes lobe steering possible.

The matter of device spacing poses another set of challenges. The smaller the spacing can be made relative to wavelength, the better a line array can approximate the behavior of a continuous radiator. When device spacing becomes large relative to a wavelength—roughly in the range of a full wavelength—the off-axis response of the array will contain many lobes and nulls. It is likely that one or more of these off-axis lobes will approach the level of the on-axis radiation. When one considers the small wavelengths of the higher audible frequencies—the wavelength of 10 kHz is 34.4 mm (1.35 inch)—the challenge of achieving optimal device spacing for higher frequencies becomes apparent. The continued reduction in size of motor assemblies through the use of high-powered magnetic materials has been helpful in addressing this issue.

2.8.5 Crossovers

Multiway loudspeakers incorporate a crossover network. A crossover network is a collection of electrical filters, each of which allows a specific portion of the frequency spectrum to pass through it. The filtered signal is then applied to one of the bands in the loudspeaker. The types of electrical filters used to execute the crossover function are low pass, high pass, and bandpass.

The simplest crossover network consists of a low-pass and a high-pass filter for use in a two-way loudspeaker. Choices that must be made regarding the filters in this crossover are:

1. Crossover frequency: below this frequency, output from the low-frequency section (woofer) is dominant, and above it the high-frequency section (tweeter) dominates.

2. Filter slopes: analog filters have characteristic stopband, or rolloff, slopes, which are integer multiples of 6 dB/octave (or equivalently 10 dB/decade). The simplest type of filter is the first order, or 6 dB/octave filter. In passive loudspeakers, the highest order filters in common use are third-order (18 dB/octave), whereas fourth-order (24 dB/octave) Linkwitz-Reilly filters are popular in active crossover implementations.

2.8.5.1 Effect on Maximum Output

The choice of filter slopes used in a crossover has a number of implications for the performance of the loudspeaker system. Generally speaking, crossover filter characteristics will affect a loudspeaker's maximum output capacity, amplitude and phase response, and directivity.

Since all transducers have a maximum excursion beyond which their output is no longer linear (or permanent damage occurs), and since the required excursion for a given acoustic output level increases with decreasing frequency, the characteristics of the high-pass filter(s) in a crossover have a direct bearing on a loudspeaker's maximum available acoustic output: in general, selecting a higher cutoff frequency will reduce the excursion required of the high-frequency transducer(s), as will employing steeper filter slopes. For a given high-frequency transducer, increasing the crossover frequency reduces the displacement required of that transducer. The demand made of the woofer as a result of the increase is strictly thermal, since the lower end of its band of use is not affected by such a change. This benefit has to be balanced against the possible inability of the woofer to effectively radiate higher frequencies over a large angle.

In addition to excursion limiting, the bandwidth of the signal applied to a given transducer determines the thermal load the transducer will see in operation. For this reason, dividing the spectrum into a greater number of bands—thereby reducing the total power that is applied to any single band—can also increase the available acoustic output of a loudspeaker. One must consider, however, that very seldom will the signal applied to a loudspeaker contain a constant broadband spectrum. At times, much of the energy applied to a loudspeaker may be confined to a relatively narrow range of frequencies. In such cases, the advantage of having a greater number of loudspeaker bands is substantially reduced.

2.8.5.2 Effect on Loudspeaker Response

The choice of filter slopes and alignments has major implications for the response of a multiway loudspeaker. Even though these effects have been examined and published for decades, they are often either misunderstood or simply ignored by loudspeaker designers.

It is a good idea to state as simply as possible the ideal functional requirements that should be met by a crossover network: a crossover should enable the acoustic sum of the individual transducers' outputs to be an accurate replica of the system input signal.

The response of a loudspeaker, for purposes of this chapter, is defined as its pressure response at a particular point in space. Even though the above criterion is simple to state, there are many design constraints that lead to tradeoffs in a loudspeaker's accuracy. For example, prevention of damage to transducers is often an overriding consideration in the design of a crossover. This may motivate the designer to consider steeper filter slopes. In some loudspeaker configurations, off-axis response anomalies are intrinsic to the design. The designer may wish to make off-axis anomalies in amplitude response as geometrically symmetrical and as narrowband as possible. The Linkwitz-Reilly filter family is sometimes employed in pursuit of these goals.

The simplest crossover is a first-order filter pair. In a two-way loudspeaker, the first-order transfer functions for low-pass and high-pass functions are:

$$F_1 = \frac{\omega_0}{S + \omega_0} \tag{2.7}$$

and

$$F_h = \frac{\omega_0}{S + \omega_0} \tag{2.8}$$

where,
F_1 is the low-pass transfer function,
$\omega_0 = 2\pi f_0$ is the angular cutoff ($-3\,\text{dB}$) frequency,
S is the Laplace complex frequency variable,
F_h is the high-pass transfer function.

If we add the two electrical transfer functions, we get

$$T_t = \frac{S + \omega_0}{S + \omega_0} \tag{2.9}$$
$$= 1$$

The two transfer functions add up to a constant, independent of frequency. This is a desirable result, since the outputs of the radiators in a multiway loudspeaker are ultimately recombined by (acoustic) addition. The transfer function of our electrical sum implies that, in a two-way loudspeaker with ideal, perfectly coincident transducers and a first-order crossover, the system transfer function would not depend on frequency. We could, with some additional effort, engage in the same exercise with higher-order transfer functions. If we did so, we would find that, of all symmetrical (identical low-pass and high-pass slope and alignment class) filters, only the first-order pair does not introduce phase or amplitude error or both to the loudspeaker's transfer function. The interested reader will find detailed mathematical analyses of the various crossover topologies in the references cited at the end of this chapter.

One way of examining the effects of crossover filters on loudspeaker response is to use circuit simulations to model various aspects of the system's behavior. This method has the advantage of presenting a simple graphic representation of the values being modeled without requiring extensive mathematical skills for comprehension.

2.8.5.3 Two-Way Crossovers

For simplicity, we will examine several aspects of crossover performance in two-way systems. Then we will point out some of the elements that must be altered when three- or four-way systems are contemplated.

The chart in Fig. 2.46 is the impulse response of a first-order crossover, including the input signal, low-pass, high-pass, and summed signals. For simplicity, the crossover frequency has been set at 1 kHz. The choice of crossover frequency causes no loss of generality.

Note that, although low-pass filter has obvious delay and the high-pass filter overshoots the input signal's return to zero, these effects perfectly cancel each other, rendering the input and the summed signals identical. This characteristic is unique to a family of crossovers identified by Richard Small as "constant voltage crossovers." The first-order filter set is the only symmetric low-pass/high-pass filter pair that falls into this class.

By contrast, the summed second-order impulse response shown in Fig. 2.47 contains significant deviations from ideal.

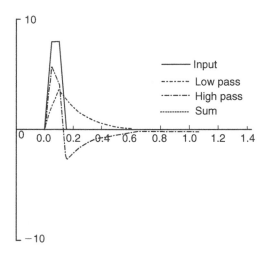

FIGURE 2.46   Impulse response family of first-order crossover.

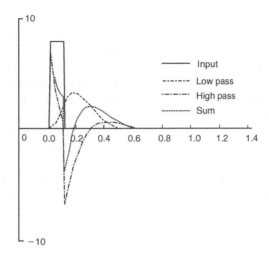

FIGURE 2.47   Impulse response family of second-order crossover.

Note that, in the summed signal (output) there is overshoot on the return to zero, followed by a delayed reaction due to the low-pass filter's delay characteristics. Viewed in the frequency domain, the second-order summed low-pass and high-pass response has a perfect null—i.e., a notch that is infinitely deep on a decibel scale—at the crossover frequency. Its phase response goes through a wrap of 360° centered at the crossover frequency.

Figure 2.48 shows the impulse response family of a fourth-order Linkwitz-Reilly filter pair. It should be evident from this series of graphs that the impulse response of a loudspeaker may be compromised by the designer's choice of crossover filter topologies. Viewed in the frequency domain, the Linkwitz-Reilly filter pair exhibits ideal amplitude response (i.e., perfectly flat) through the crossover range and elsewhere, but its phase response goes through a 720°

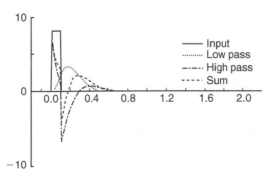

FIGURE 2.48   Impulse response family of fourth-order Linkwitz-Reilly filters.

wrap through the crossover region. From what we have observed so far, it is evident that higher-order symmetric filters can introduce nonideal transient response behavior when used as crossover filters. Based on observations of the delaying effect of the low-pass filters, one might be tempted to introduce electrical delay into the high-frequency signal in an attempt to better synchronize low- and high-frequency signals. In the case of the LR filter family, such attempts will only serve to compromise the amplitude response of the loudspeaker while offering minimal improvement in the impulse response. As with crossover-frequency anomalies caused by noncoincident transducers, one way of addressing the nonideal behavior of higher-order symmetric filters is to assert that the problems are not audible. It is also possible to address transient response issues and at the same time retain a steep filter slope for one of each pair of neighboring bands in a multiway loudspeaker. Crossovers of this type are termed *constant voltage* crossovers and are discussed in Section 2.8.5.3. The simulations above are based on ideal filter behavior and ideal transducers. As one makes the simulation more realistic, accounting for the bandpass behavior of real-world transducers, the performance of all of the modeled crossovers will deteriorate, but the relative attributes of constant voltage filters remain.

2.8.5.4 Beyond Two-Way Systems

As the number of spectral bands in a loudspeaker increases, the issues that must be dealt with in crossover design multiply. In a system with three or more bands, at least one of the crossover filters is a bandpass, usually formed by cascading low-pass and high-pass filters of the desired characteristics. The low-pass portion of the arrangement will introduce delay in its passband, which can create misalignment between the band in question and its lower neighbor. In addition to this issue, there is also the possibility of interactions between transducers that are not neighbors in the audio spectrum (e.g., the woofer in a three-way system can contribute enough energy in the high-frequency horn's pass band to make its presence known). This type of interaction is often undesirable, as it has generally deleterious effects on the response and directivity of the system.

2.8.5.5 Passive versus Active Crossovers

When designing a passive crossover—one that receives the power amplifier's output and applies appropriately filtered signals to each transducer—the designer must account for the frequency dependence of the impedances of each transducer in the system. In the case of most cone transducers, the impedance curve has a peak at the resonant frequency, above which it decreases to a minimum and then rises with frequency in similar fashion to the impedance of an inductor. This variation of impedance with frequency is often minimized through the use of a parallel, or shunt, network. Once the device's impedance has been stabilized in this manner, the actual crossover filter may be designed to drive a purely resistive load with excellent results. Active crossovers—those that divide the spectrum at line level and apply the band signals to the inputs of power amplifiers—have the advantage of the buffering effect provided by the power amplifier. Impedance-related issues are far less significant in this case, and active filters—particularly DSP-based ones—offer a number of options not readily available in passive versions. These include frequency-independent delay, all-pass filters, and dynamics processing (compression/limiting). The price that is paid in an active system is in additional channels of power amplification and wiring.

2.8.6 Acoustic Boundaries

Generally, one considers that acoustic boundaries are part of the space into which a loudspeaker is radiating. The field of architectural acoustics is largely concerned with the acoustic behaviors such boundaries cause. However, every loudspeaker has a collection of acoustic boundaries independent of the external environment in which it is operated, and these boundaries make a surprisingly large contribution to the loudspeaker's response and directivity.

Most of the boundaries associated with loudspeakers constitute reflective surfaces: enclosure walls are designed to be rigid and generally have hard surfaces. The same is true for horn surfaces. Phenomena associated with this type of surface fall into two broad categories: reflection and diffraction.

In the simplified textbook models such as a piston radiating into a half space, the infinitely large baffle on which the loudspeaker is mounted is assumed to be perfectly reflective. All of the reflections that occur at this surface will add coherently to the outgoing wave, since the source is in the same plane as the baffle. The only interfering radiation present in this model is that which is caused by the source itself, and it is this simplicity that allows a closed form solution—the piston directivity function—to yield an accurate prediction of the device's behavior.

If a hard surface is present on the front of the baffle and at right angles to it—as would be the case with room walls, for example—the wave's outgoing motion can continue no farther past this surface. Its direction is reversed due to reflection.

In a typical direct radiator loudspeaker, the wave created by a transducer expands along the front surface of the cabinet until it reaches the edges. At these edges, the support provided by the enclosure's front surface for forward motion of the wave abruptly collapses as the wave is allowed to expand rearward as well as forward. The propagation of the sound wave past this point is altered by diffraction.

Loudspeaker cabinet diffraction has not been a well-understood phenomenon until relatively recent work. The model developed by Vanderkooy shows that diffraction at an edge has strong dependence on the observation angle and that forward diffraction (in the same direction as the original outgoing wave) is inverted in polarity, whereas diffraction at angles greater than 180° (to the rear of the loudspeaker) is of the same polarity. The reader is encouraged to study Vanderkooy's work, as well as the other references, for mathematical treatments of this phenomenon.

The net effect of this diffracted energy is to introduce a set of acoustic arrivals at an observation point that follow the direct arrival in time and are reversed in polarity for positions in front of the loudspeaker. These arrivals interfere with the direct signal, with the specific effect of the interference depending on frequency, baffle size, and transducer positioning on the baffle. The result is a series of peaks and dips in the loudspeaker's response due entirely to the baffle itself.

Some effects of diffraction from panel edges are illustrated in the following graphs. On-axis response measurements were performed on a 1 inch soft dome tweeter with a 3.75 inches (95 mm) square mounting panel. Figure 2.49 is a response measurement of the tweeter alone, suspended from a microphone stand. Figure 2.50 is the same tweeter mounted on a thin panel approximately 19 inches (483 mm) square.

Note the relatively wide depression in the tweeter's response in Fig. 2.49. The center of the depression is approximately at 6.5 kHz. A diffracted arrival at a one-wavelength distance will interfere destructively with the primary wave. At 6.5 kHz, this distance is approximately 2.1 inches. This is consistent with the average distance from the center of the tweeter mounting flange to its edge.

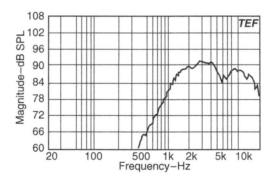

FIGURE 2.49   Dome tweeter on axis with no baffle.

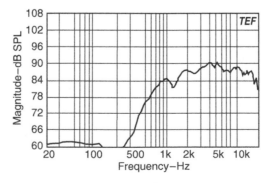

FIGURE 2.50   Dome tweeter on axis with 19 inch square baffle.

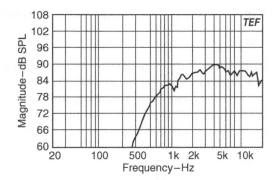

FIGURE 2.51 Dome tweeter on axis, mounted on 19 inch square baffle with absorption on its edge.

A tweeter with a round mounting flange could be expected to have a deeper, narrower notch due to reduced time smear in the diffracted arrival.

The same characteristic notch is present in Fig. 2.50, but at a much lower frequency. This is also consistent with the model of reversed-polarity forward diffraction: The notch is now centered at 1220 Hz, which has a wavelength of approximately 11 inches. The average distance from the center of the 19 inch panel to its edge matches this dimension very closely.

Figure 2.51 is the same configuration as in Fig. 2.50, with the addition of a layer of ¾ inches (19 mm) thick foam attached at the edges of the panel. This material is relatively absorptive above 1 kHz. Its effect on the tweeter's response is most evident between 1 kHz and 3 kHz. The graphs in Figs. 2.50 and 2.51 display loudspeaker response differences that are due entirely to the boundaries formed by the speaker's baffle. The same transducer was used in each measurement.

This brief examination of some acoustic effects due to loudspeaker boundaries is intended to provide a starting point for further study and investigation. A number of implications for loudspeaker design should be readily apparent.

Transitional points in a loudspeaker's shape (e.g., edges, slots) behave as acoustic sources. Energy arrivals from these features always follow the primary wave in time. Additionally, they can be reversed in polarity. Acoustic absorption is a useful diagnostic tool as well as a powerful design element for the loudspeaker engineer.

2.8.7 Conclusion

Loudspeaker system performance is a function of several elements, including transducer design, crossover topology, component location and orientation, and the acoustic boundaries formed by the loudspeaker's housing. Each of these elements has a major effect on the final result, and the most effective loudspeaker designs successfully address all of these areas.

2.9 CHARACTERIZATION OF LOUDSPEAKER PERFORMANCE

2.9.1 Motivation

In considering the behavior of a loudspeaker, it stands to reason that we need performance parameters with which we can evaluate the effectiveness of a specific device for an envisioned use. There are many performance areas in which loudspeakers differ in significant ways, including on-axis response, bandwidth, directivity, distortion, and maximum acoustic output. Unfortunately, there are a number of different formats for presentation of loudspeaker performance data. Before attempting to interpret such data, it behooves us to develop some general concepts of loudspeaker performance.

The picture is made much more complicated by the fact that we hear not only the direct sound produced by a loudspeaker, but also the reflections caused by interactions between the loudspeaker and the acoustic environment in which we are listening. Different loudspeakers interact in different ways with acoustic environments, with certain types and degrees of interaction being preferable to others. For this reason, it is useful to develop a concept of loudspeaker performance that will provide a means for understanding (and, hopefully, predicting) these interactions.

2.9.2 Efficiency and Sensitivity

Since a loudspeaker converts electrical energy into acoustic energy, the concept of efficiency is relevant. As we will see, this conceptual construct, while it is a good starting point for study, has limitations in the characterization of loudspeaker performance.

Efficiency is defined as the ratio of power provided by the system output divided by the power applied to the input. As a result of conservation of energy, the efficiency of a loudspeaker (or any energy-conversion device) is always less than one. Most often, efficiency is expressed as a percentage. Typical loudspeaker efficiencies range from less than 1% in the case of some hi-fi products to approximately 25% for limited-bandwidth horn-loaded devices.

Since a loudspeaker's efficiency varies with frequency, a single number for efficiency does not generally provide adequate information for discriminating one device from another. Also, since human hearing responds to changes in acoustic pressure, the total power radiated into an acoustic space may or may not be a good indicator of what the human ear–brain perceives. Furthermore, the devices that drive loudspeakers—incorrectly called *power amplifiers*—are designed to control the voltage applied to a loudspeaker. For these reasons, the concept of a loudspeaker's functional efficiency needs to be expanded.

The parameter most often used to characterize a loudspeaker's ability to produce acoustic output is called *sensitivity*. A loudspeaker's sensitivity is the sound pressure level (SPL) produced at a reference distance with a reference electrical input signal. The most common standard is dB-SPL at 1 meter with a

1-watt input. Since a loudspeaker's impedance varies with frequency, and since a power amplifier is actually a voltage-controlled voltage source, the 1 watt figure is usually translated into an rms voltage (e.g., 2.83 Vrms into $8\,\Omega = 1\,W$). Also noteworthy is the fact that the actual measurement will generally not be accurate if it is actually carried out at a 1 meter distance, since it will not be in the loudspeaker's farfield. Instead, the testing is done at a greater distance and the results normalized to the 1 meter reference distance. The discussion that follows is based on the premise that we are interested in knowing the acoustic output characteristics of a loudspeaker with known voltages applied to its input.

2.9.3 Network Transfer Function

The performance characterization of electrical circuits is a well-developed realm and is employed as the basis for much data presented in regard to loudspeaker behavior. The impulse response, $H(t)$, and LaPlace transfer function, $L(H[t]) = F(s)$, of an electrical circuit are widely used models for the linear portion of a circuit's behavior. A large body of practical mathematics has been developed to aid in manipulating transfer functions of circuits, and the subject is covered at the undergraduate level in almost every engineering discipline.

A necessary item in applying these concepts to an electrical network is a definition of which terminals constitute the input and which will be considered the output. Using these definitions, the performance of the circuit may be modeled and/or measured and the resulting data used to evaluate the suitability of the circuit for an intended use.

2.9.4 Loudspeaker Transfer Function

Before developing these concepts further, we should recognize the importance of the correlation between the response behavior of a loudspeaker and its audible (i.e., subjectively evident) performance. The inevitable limitations in the resolution of measured loudspeaker data should ideally be determined with the capabilities and limitations of human hearing in mind. Data that is too highly resolved will reveal a number of details, or artifacts, that are not likely to be audibly significant, while insufficiently resolved data will tend to smooth over relatively serious imperfections that may easily be heard. With respect to frequency resolution, constant percentage octave (log frequency) resolution would appear to correlate best with the capabilities of human hearing. If measurements are taken so as to yield $\frac{1}{6}$-octave resolution above 100 Hz, it is unlikely that greater resolution would reveal additional features that can be distinguished by human hearing.

The concept of frequency response (more appropriately, amplitude response) is a direct consequence of the transfer function model. This is the most familiar of the many possible ways of graphically representing portions of a transfer function. It is widely assumed that a loudspeaker may be characterized by one frequency response, usually measured at a point defined to be on axis of the loudspeaker. This assumption is incomplete: a loudspeaker has infinity of transfer functions (or, interchangeably, impulse responses), one for each point in 3D space. In the interest of compactness, we could say equivalently that a loudspeaker has a transfer function with four independent variables instead of one: where $F(s)$ is sufficient for electrical circuits, for loudspeakers the equivalent expression (in Cartesian coordinates) will be $F(s,x,y,z)$.

When we consider loudspeakers, the analogy with electrical networks is incomplete due to the nature of the device's output. Whereas a two-port electrical network has a single output, a loudspeaker radiates energy into free space in all directions. If only one listener were present, and if there were also no reflections in the acoustic environment, then the loudspeaker's response at a single point—the listening location—would be sufficient to characterize what that listener would hear. If multiple listeners and reflections are present, it is no longer sufficient to consider only the single transfer function: we need much more information.

If we limit our consideration to the far field (i.e., distances many times greater than the largest dimension of the loudspeaker), then the dependence of the transfer function on distance will be reduced, in most practical cases, to a characteristic delay of

$$\tau = \frac{r}{c} \tag{2.10}$$

where,

r is the distance from the source to the observation point,

c is the phase velocity of sound in air, plus a change in acoustic pressure that is inversely proportional to distance from the source due to the inverse square law.

Both of these quantities may be assumed to be frequency-independent, although there are some exceptions.

Given this simplification, the extended transfer function of a loudspeaker may be characterized on the surface of a sphere of some arbitrarily chosen radius with the source at its center. The number of independent variables is then reduced to two, yielding a transfer function that may be represented as a function of S and two angles—i.e., $F(s,\theta,\psi)$. Even with this simplification, the number of single-point transfer functions remains uncountably infinite. Clearly, further simplification will be required if the task of measuring and describing a loudspeaker's performance is to be made practically realizable.

Currently available software for the simulation of sound system performance requires data to be presented with a fixed angular resolution. The polar coordinate system that has been adopted for this purpose is most easily described as that of a globe with the loudspeaker's axis aimed at the North pole. Typically, the plane of horizontal coverage is defined as 0° rotation angle, with the lines of constant rotation equivalent to longitude and radius angles analogous to latitude. One advantage of this approach is that data points are at maximum density near the on-axis position. There is still debate regarding the angular resolution required to show relevant details of device performance. Increments as fine as 1° have been suggested. Practically speaking, even with 10° increments, a complete set of measurements on a device with mirror-image symmetry (i.e., requiring measurements in only one quadrant) requires 172 response measurements of the device. An asymmetric device (e.g., Altec VIR, IMAX® PPS, requiring two quadrants of measurement) requires 325 measurements to characterize with 10° resolution.

One possible compromise is to use one angular increment for measurements taken within the intended coverage pattern of the loudspeaker and another, broader, one for other measurements. This has the advantage of providing greater detail in the angular area in which the loudspeaker's response has the greatest audible effect. It would, however, complicate the process of interpolation that is required to approximate the response of a speaker at angles that fall between the angles at which measurements were taken. This variable resolution is unavailable in current array prediction software.

Due to the large amount of measured data that is required to meaningfully characterize the performance of a loudspeaker, it is highly impractical, if not altogether impossible, to provide the data in a hardcopy format. In order to conveniently view loudspeaker data of this complexity, one must employ a computer program. For many years, the only available programs for this purpose were those that were primarily designed for sound system modeling and prediction. These programs have capabilities that go far beyond the display of loudspeaker data, are not optimized for that use, and are typically quite costly.

Recently, a format specifically for presentation of loudspeaker performance data, called the *common loudspeaker format*, or CLF, has been developed. This format is supported by a consortium of loudspeaker manufacturers. Due to the amount of data accommodated by the format, it is optimized for electronic, rather than hardcopy, presentation of data. It requires a data viewer program, which is available for download free at http://clfgroup.org. The displays in the CLF viewer include 3D amplitude balloons, traditional polar plots, normalized off-axis response plots, impedance versus frequency, as well as other data. Figures 2.52 and 2.53 are screen captures of a CLF display.

2.9.5 Impedance

The impedance of a loudspeaker is very seldom constant with respect to frequency. For this reason, the nominal impedance provided in the specification sheet—typically $4\,\Omega$, $8\,\Omega$, or $16\,\Omega$—is often useless as a figure of merit. Because power amplifiers have limited ability to drive excessively low impedances and because loudspeaker cabling may have nonnegligible series resistance, it behooves the prospective purchaser of a loudspeaker to examine its impedance versus frequency curve. Of greatest interest is the minimum impedance seen in

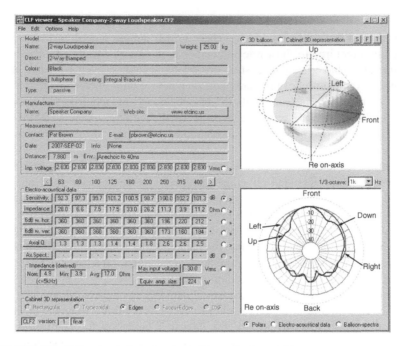

FIGURE 2.52   CLF main window, showing balloon and polar graphics.

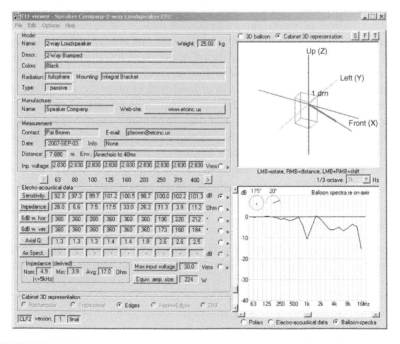

FIGURE 2.53   CLF main window, showing cabinet drawing and normalized off-axis response.

the device's bandwidth and, to a lesser extent, the frequencies and magnitudes of any peaks in the curve.

2.9.6 Distortion

The concept of a transfer function assumes a linear system. In a linear system excited by a single frequency, the output will contain only that frequency, possibly changed in amplitude and/or phase. By extension, the output of a linear system excited by a signal containing multiple frequencies will contain only those frequencies present in the original signal.

There are a number of nonlinear mechanisms in any loudspeaker. These include nonlinearities in the motor, suspension, and air (e.g., in a phasing plug or the throat of a horn). For this reason, all practical loudspeakers have nonnegligible levels of harmonic and intermodulation distortion. By comparison with modern electronic signal processing devices and amplification, loudspeakers have orders of magnitude higher levels of distortion.

The degree to which loudspeaker distortion constitutes an audible problem is a matter of some controversy. Indeed, some of the more popular devices have distortion levels that are quite high, even by loudspeaker standards. Various studies over the years have established the audibility of simple harmonic distortion at levels above approximately 2%, but it is not clear that all of these studies fully accounted for the distortion present in the loudspeakers required to perform the testing. The continuing popularity of tube amplifiers among the audiophile community would tend to indicate that at least some forms of distortion are perceived as pleasing. In terms of simple harmonic distortion, it is often asserted that even-order distortion products have a more musical relationship to the fundamental, going upward in frequency in successive octave steps. Because of this, it may be true that even-order distortion products are more readily tolerated (or even preferred) by many listeners.

Because of the relatively high levels of distortion in all loudspeakers and the wide variety of ways in which a signal can be distorted, there is no consensus in the industry as to the best means for characterizing the distortion performance of a loudspeaker. The audible significance of the nonlinear distortion caused by a loudspeaker is best judged in person, and the results may or may not correlate well with commonly measured data.

2.9.7 Characterization for Design Purposes

Prior to beginning work on a new design, the designer will (or should) develop a set of performance criteria for a loudspeaker. Parameters typically considered are bandwidth, available acoustic output, directivity, and efficiency. Often, the designer must interpret subjective information provided by others and translate it into performance specifications.

The more complete the data regarding target performance, the more satisfactory the finished loudspeaker is likely to be. For this reason, targeted applications for a new design should be well understood and well defined. Possible performance definitions include characteristic isobars, lower and upper cutoff frequencies, tolerance for nonideal amplitude response (off-axis as well as on-axis), maximum acoustic output, maximum distortion level relative to fundamental, and phase versus frequency criteria. A reasonably good idea of minimum acceptable performance is also useful for purposes of cost engineering. Once the overall loudspeaker performance envelope is finalized, requirements for individual component performance can be established.

The specific measurements that should be performed on an individual component will depend on the transducer or radiator in question and on the nature of the loudspeaker of which it will become a part. Impedance versus frequency measurements are always essential. For woofers, this will allow determination of the parameters necessary for the design of an appropriate enclosure. For horn/driver combinations, the designer needs to know the frequency of mechanical resonance. It is also possible to identify internal horn reflections and diaphragm breakup problems in an impedance curve. Finally, the behavior of the component as an electrical load is required for the design of passive crossovers.

Measuring a representative set of transfer functions is an essential part of the component characterization process. What constitutes a representative set will depend on the component. A woofer will need comparatively few measurements if it is well behaved and to be used only at low frequencies, a horn will need a significantly larger number of measurements, and an array of two or more components operating over the same range of frequencies will require still more measurements. Some devices cannot be suitably characterized by a reasonable number of measurements. The measurement process will determine if a component is usable in the intended application and will aid in early prediction of the ultimate performance of the completed loudspeaker.

The extent of testing needed to evaluate the most complex component is likely to be required for the complete system. The degree to which target performance objectives have been met should be established at the prototype stage. Any necessary modifications may then be made and the system retested. This process may continue through as many iterations as necessary for the loudspeaker to perform as desired.

2.9.8 Characterization for the User

Once a loudspeaker is in production, performance data will be required for (a) giving potential buyers information for comparative purposes, and (b) use in the design of sound systems. Loudspeaker performance data provided to the sound system designer should be sufficiently comprehensive for acceptably accurate prediction of the performance of a sound system. Unfortunately, the volume of data required to fully characterize a loudspeaker is, as we have discussed, quite large. If hard copy were generated with all pertinent information, most loudspeakers intended for professional use would require a small book.

There are many ways to provide transfer function information about a loudspeaker. Keeping in mind that our extended definition of *transfer function* for a loudspeaker intrinsically includes directivity information, possible formats include:

1. Amplitude response curves calibrated to a constant level reference (e.g., dB-SPL) with a specified signal input (e.g., 2.83 Vrms) at a variety of angles. This format has the advantage of explicitly showing the direct-field response that listeners in various locations relative to the loudspeaker will hear, Fig. 2.54.
2. Amplitude response measurements as above but normalized to the response at a particular angle, usually on axis. This is the equivalent of assuming that the on-axis response will be equalized flat, which is not always a good idea. The caveat: anomalous narrowband behavior on axis (i.e., a notch) that disappears off axis will create features (peaks) in the normalized curves that are unrepresentative of what would occur in actual use.
3. On-axis amplitude response, accompanied by polar plots at various frequencies. The common usage of only vertical and horizontal polar curves is problematic. The omission of polar curves at angles between 0° and 90° rotation leaves a lot of a loudspeaker's performance to speculation, Fig. 2.55.
4. On-axis amplitude response, accompanied by isobars at various frequencies. This format is useful for showing overall coverage behavior of a loudspeaker. Lobes, where present, are not generally revealed in isobar plots.
5. Directivity data for use with sound system modeling software. The format of this data will be dictated by the requirements of the predictive program. Some standards for the presentation of this are being discussed, but there is not at yet an industry-wide consensus on a final standard, Figs. 2.56 and 2.57.

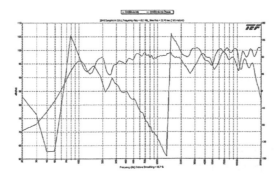

FIGURE 2.54 Graph of a loudspeaker's on-axis amplitude and phase response, $^1/_6$-octave smoothed. Courtesy Frazier Loudspeakers.

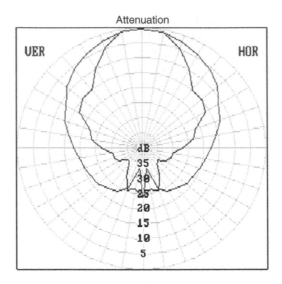

FIGURE 2.55   Vertical and horizontal polars, one octave averaged. Courtesy Frazier Loudspeakers.

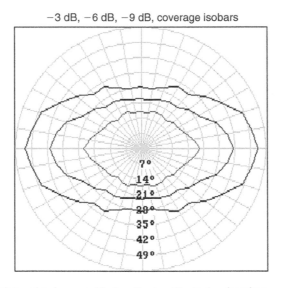

FIGURE 2.56   Octave band averaged isobar. Courtesy Frazier Loudspeakers.

Of course, various combinations of the above can be provided. The formats available for the presentation of loudspeaker data continue to evolve. The availability of inexpensive mass data storage media and ever more sophisticated acoustic modeling software will continue to make the presentation of loudspeaker response and directivity information more effective and intuitive.

3D directivity pattern

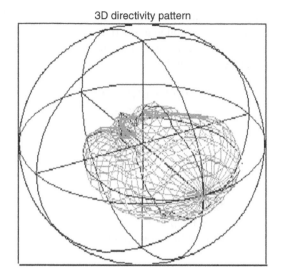

FIGURE 2.57   Three-dimensional representation of octave band isobar. Courtesy Frazier Loudspeakers.

2.10 DIRECT RADIATION OF SOUND

The physics and mathematics of loudspeaker behavior are diverse and complex. In order to account for the conversion of an electrical signal to sound, one must develop both acoustic and electromechanical models. Several of these models, which are developed and presented in almost every introductory text on acoustics, are presented here without proof. The interested reader is encouraged to study the references.

2.10.1 Acoustics of Radiators

An understanding of direct sound radiation from a piston in space, a baffle, or a box can be approached by analyzing two distinct but directly related quantities, radiation resistance and directivity. Radiation resistance is the measurement of the capacity of an acoustic radiator to convert vibratory motion into sound energy. It is the ratio of pressure to the volume velocity due to the piston's motion. At high frequencies, all pistons have the same capacity per unit surface area to produce acoustic power. However, as the size of the wavelength of sound being produced approaches the size of the piston, the radiation resistance decreases as the square of frequency, i.e., at approximately 12 dB/octave.

2.10.1.1 Piston in an Infinite Baffle

A piston in a wall of infinite extent (half space) is the model most commonly employed to develop predictive equations. Even though this model is not representative of the majority of actual loudspeakers, its simplicity and mathematical manageability make it useful for instructional and comparative purposes.

A piston in an infinite baffle will see an acoustic load that depends on its size relative to a wavelength of sound at the frequency of interest. The radiation resistance, which is the part of the acoustic impedance that accounts for transmission of sound energy, is given by

$$R_f = \rho_0\, c\pi a^2 [R_1(2ka)]$$
$$= \rho_0\, cS\, [R_1(2ka)] \tag{2.11}$$

where,
ρ_0 is the equilibrium density of air,
c is the velocity of sound in air,
a is the radius of the piston,
k is the wave number, $2\pi f/c$,
$S = \pi a^2$ is the surface area of the piston,
$R_1[2ka]$ is the piston resistance function, given by

$$R_1(x) = \frac{x^2}{2 \bullet 4} - \frac{x^4}{2 \bullet 4^2 \bullet 6} + \frac{x^6}{2 \bullet 4^2 \bullet 6^2 \bullet 8} - \ldots \tag{2.12}$$

The value of the piston resistance function approaches unity for values of $2ka$ above 6. For example, in the case of a piston with an effective radius of 6 inches, the radiation resistance will be approximately constant above 1100 Hz.

The acoustic power radiated by a flat piston is given by

$$W = \frac{R_r U_0^2}{2}$$
$$= \frac{U_0^2 \rho_0\, c\pi a^2 R_1[2ka]}{2} \tag{2.13}$$
$$= U_0^2 \rho_0\, cS R_1[2ka]$$

where,
U_0 is the amplitude of the piston's velocity.

Two regimes of interest may be derived from the above equation. If we first consider $2ka < 1$—i.e., a small piston and/or low frequency—we can neglect the higher-order terms in the expression for the piston resistance function

$$R_1(x) \approx \frac{x^2}{8} \tag{2.14}$$

and the power radiated by a flat piston becomes

$$W = \frac{\rho_0\, ck^2}{4\pi}(S^2 U_0^2) \tag{2.15}$$

Note that, for constant velocity amplitude, the acoustic power rises as the square of the frequency. Clearly, there must be a compensating mechanism, as a typical cone transducer has relatively flat amplitude response over this range of frequencies. This mechanism is the mechanical impedance due to the moving mass of the piston, which rises with the square of frequency. Therefore, a piston excited by a force that does not vary with frequency responds with a velocity that falls off as the square of frequency. But, in the low-frequency region, the acoustic impedance rises with the square of frequency, so the two effects effectively cancel each other over a significant range of frequencies. It is this serendipitous balance between key mechanical and acoustic parameters that makes the cone transducer an effective acoustic radiator.

The second regime is the region for which $ka \gg 1$ (high frequencies and/or a large piston). In this case, because the piston resistance function approaches unity, we get

$$
\begin{aligned}
W &\cong \frac{1}{2}\rho_0 c\pi a^2 U_0^2 \\
&= \frac{1}{2}\rho_0 cSU_0^2.
\end{aligned}
\tag{2.16}
$$

Note that there is no frequency dependency in the above expression: the radiation resistance of a piston approaches a constant at high frequencies. Given velocity amplitude that falls off as the square of frequency, it is clear that, in the high-frequency regime, the acoustic power radiated by a typical transducer could be expected to decrease as the fourth power of the frequency. Note also that, in the low-frequency limit, for constant velocity amplitude, radiated power goes as the square of the surface area of the piston. In the high-frequency limit, however, it goes as only the first power of the area. Therefore, all else being equal, increasing the size of the piston has a greater effect on its low-frequency output than on its capacity to radiate higher frequencies.

2.10.1.2 Piston Directivity

So far, we have examined expressions for the total power radiated by a piston. If a piston radiated identically in all directions, no further acoustic information would be needed. Since this is not the case, it is also worthwhile to consider the nature of this directivity.

The mathematical technique for deriving the piston directivity function is to consider the piston as being made up of infinitesimal differential elements, each of which contributes to the observed radiation at a point in space. These individual contributions are combined via integration to yield a value for each specific point in space.

In coming up with a manageable expression for piston directivity, one assumption must be made: the distance from the piston to the observation point is much greater than the piston's radius. The result for the pressure amplitude is

$$\rho = \frac{\rho_0\, ckaU_0}{2r}\left[\frac{2J_1\,(ka\sin\theta)}{ka\sin\theta}\right] \tag{2.17}$$

The first term in the above relationship contains the dependency of the pressure on velocity amplitude, piston size, and distance from the source. The second term, called the *piston directivity function*, is derived from a Bessel function, $J_1(x)$. The value of this function is graphed in Fig. 2.58. Note that, up to $ka = 3.83$, the value of the piston directivity function is uniformly positive. The radiation pattern of the piston will have only a single lobe under these conditions. If $ka = 3.83$, the pattern will have a null at 90° off-axis. For higher values of ka, this null will occur at successively smaller angles. Additionally, secondary lobes will appear outside of the main lobe, although these lobes are smaller in magnitude than the primary one. These lobes will alternate in sign: the first set will be negative, the second positive, etc.

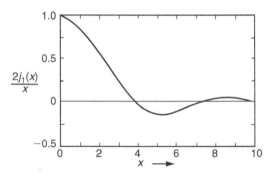

FIGURE 2.58   Piston directivity function.

The directivity of a real loudspeaker differs from that predicted for a rigid piston due to the fact that several of the basic assumptions in the preceding model are not fully satisfied. First, no real loudspeaker has a perfectly rigid cone or diaphragm. In the case of a cone transducer, the diaphragm is excited at its center. The excitation travels outward from the voice coil as an acoustic disturbance in the cone material. The velocity of propagation of this disturbance is always finite. At lower frequencies, this effect is negligible, but at higher frequencies not all portions of the diaphragm will vibrate in phase.

A second difference between real loudspeakers and our theoretical piston is that practical diaphragms are very seldom flat. Most often, they are in the shape of a concave cone, but convex dome shapes are also employed. In many instances, the shape of the diaphragm is chosen so as to minimize the effect of finite-velocity wave propagation in the diaphragm material on the device's on-axis response.

Generally speaking, the directivity of real-world cone or dome transducers is qualitatively similar to that of a rigid, flat piston. The nonideal behavior of real transducers can actually create beneficial effects in that the frequency at which secondary lobes appear can be higher than the theory predicts.

2.10.2 Direct Radiator Enclosure Design

A woofer is not effective as a freestanding radiator. If it were to be employed in this fashion, radiation from the rear of the diaphragm, which is out of phase with that from the front, would cause cancellation, particularly at low frequencies. Consequently, woofers are always housed. Two types of enclosures are widely used: sealed and vented.

2.10.2.1 Sealed-Box Systems

The low-frequency response of a sealed-box system may be modeled as a second-order high-pass filter. The effect of the enclosure is to add stiffness to the woofer suspension, which will modify the free-air resonant frequency of the woofer. The contribution made by the air in the enclosure to the stiffness of the diaphragm is given by

$$k_b = \frac{\rho_0 c^2 S_D^{\ 2}}{V_B} \tag{2.18}$$

where
k_b is the box effective spring constant,
ρ_0 is the equilibrium density of air,
c is the speed of sound in air,
S_D is the diaphragm area,
V_B is the enclosure volume.

The spring constant of the enclosure simply adds to that of the woofer suspension, so

$$k' = k_d + k_B \tag{2.19}$$

The air mass that effectively adds to the moving mass of the diaphragm is given by

$$m_a = \frac{\rho_0 8 S a}{3\pi} \tag{2.20}$$

where,
S is the surface area of diaphragm,
a is the radius of diaphragm.

Again, this mass is additive, so

$$m' = m_d + m_a. \tag{2.21}$$

Note the dependence of the enclosure spring constant and the effective mass on two properties of air: its equilibrium density and the speed of sound. Both of these quantities are subject to significant variations with atmospheric conditions, so the degree of accuracy with which one can predict the response of an enclosure/transducer combination in actual use is intrinsically limited.

The resonant frequency of the woofer/enclosure system is given by

$$\omega_0 = \sqrt{\frac{k'}{m'}} \tag{2.22}$$

where,

$\omega_0 = 2\pi f$ is the angular resonant frequency.

The expression for the low-frequency farfield pressure response of a sealed-box woofer when driven by a constant-voltage source may be written as

$$p = \left[\frac{E_m\, Bl\rho_0 S_d}{2\pi R_e m' r} \right] \left[\frac{\dfrac{-\omega}{\omega_0^{\,2}}}{1 + \dfrac{j\omega}{Q_t \omega_0} - \dfrac{\omega^2}{\omega_0^{\,2}}} \right] \tag{2.23}$$

where,

E_m is the amplitude of the applied voltage,

R_e is the dc resistance of the voice coil,

B is the flux density in the magnet gap,

l is the length of the voice coil conductor in the gap, and

r is the distance from the source to the observation point.

Q_t is given by

$$Q_t = \frac{\sqrt{k' m'}}{R_m = \dfrac{\beta^2\, l^2}{R_e}} \tag{2.24}$$

where,

R_m is the woofer mechanical resistance (damping).

Note the separation of the right side of the equation for pressure response into two parts. The first contains amplitude information resulting from the driving voltage, woofer parameters, and distance from the source, and the second provides frequency response information. A voltage excitation of the form of $E = E_m e^{j\omega t}$ is assumed.

From the first term in the equation, we can see several ways in which the system's output can be increased for a given distance and driving voltage:

1. Increase the flux density, B. Increasing magnet size will accomplish this up to the point at which the pole piece is saturated.
2. Increase the length l of the conductor in the gap. This will increase R_e, however, if all we do is to add turns to the voice coil.
3. Increase the diaphragm surface area, S_d. Doing so without changing the density of the material will also increase m, however.

Changing any of the above will potentially have an effect on the value of Q_t. If the total system Q has a value of 0.707, the response of the system will be maximally flat, also known as a *Butterworth* alignment. If the Q is higher than this, there will be a peak in the response just above the cutoff frequency.

If total Q is lower than 0.707, the response will fall off, or sag, in the region above cutoff, Fig. 2.59.

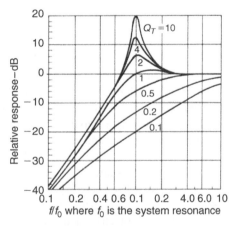

FIGURE 2.59 Response of closed-box system versus Q and normalized frequency relative to system resonance.

2.10.2.2 Vented Boxes

Prior to the existence of analytical models for a vented-box woofer, it was understood that an opening could be cut in a low-frequency enclosure, creating a Helmholtz resonance. The vent itself functions as an additional radiator in this case, and its radiation can add constructively to that of the woofer over a limited range of frequencies. A. N. Thiele developed the original published analytical model for vented box radiators, and his work was later supplemented by that of Richard Small.

The effect of the enclosure on the spring constant of the woofer is the same as in a sealed enclosure. The vent functions as a passive radiator coupled to the woofer cone via the air in the enclosure. In modeling the response of a vented enclosure woofer, we must account for the motion, and therefore the acoustic radiation, of the vent. The air in the vent is assumed to move as a unit to allow the mathematics to remain manageable. The following expression gives the far-field half-space acoustic pressure from a vented enclosure at low frequencies:

$$p = \left[\frac{E_m \beta l \rho_0 S_d}{2\pi r R_e m'_d r} \right] \times \left[\frac{\left(\frac{\omega}{\omega_0}\right)^4}{\frac{\omega}{\omega_0^4} + \frac{1}{Q_t}\left(\frac{\omega}{\omega_0}\right)^3 - \frac{\omega_v^2 + \omega_d^2}{\omega_0^2}\left(\frac{\omega}{\omega_0^2}\right) + \frac{1}{Q_t}\left(\frac{j\omega_v^2}{\omega_0^2}\right)\left(\frac{\omega}{\omega_0}\right)} \right]$$

(2.25)

where,

ω is $2\pi f$,

ω_0 is $\sqrt{(\omega_d \omega_v)}$,

ω_v is $\sqrt{(k_v / m_v)}$.

The first portion of the right side is identical to its counterpart in the sealed-box equation. The second part describes a fourth-order high-pass filter. There are three general alignment classes for such filters: Butterworth, or maximally flat; Chebychev, or peaked; and Bessel, or maximally flat group delay.

A comparison of the attributes of sealed and vented enclosures is in order. The sealed system has the advantage of an intrinsic excursion-limiting mechanism—the addition to the woofer's spring constant due to the air in the chamber—for frequencies below the system cutoff. The vented system, on the other hand, can allow excessive woofer excursion if excited with out-of-band signals, so an electrical high-pass filter is a desirable protective element. The higher-order nature of the vented system renders it more susceptible to misalignments caused by production variations in woofer parameters and changes in atmospheric conditions, but it has the advantage of requiring less woofer excursion for a given acoustic output in the lowest portion of its usable bandwidth. In general, a sealed system will have more highly damped low-frequency transient response when compared to a vented system with the same cutoff frequency. This effect is most noticeable for frequencies in the neighborhood of the system lower cutoff frequency.

The design and modeling of vented and sealed woofer enclosures has been greatly simplified in recent years due to the ready availability of computer software developed specifically for the purpose. The following Thiele-Small (in honor of A. N. Thiele and Richard Small) loudspeaker parameters are required as minimum input to an enclosure design program:

1. Q_T is the total loudspeaker Q.
2. F_S is the free-air cone resonance of the loudspeaker.

3. V_{AS} is the equivalent volume compliance of the suspension.
4. X_{max} is the maximum linear excursion.
5. P_{max} is the maximum thermal power handling.

Most manufacturers provide the above parameters as per Audio Engineering Society (AES) recommended practice on loudspeaker specifications.

Loudspeaker enclosures can have resonances, or standing waves, caused by internal reflections. The characteristic frequencies (f_b) of these standing waves are given by:

$$f_n = \frac{c}{2}\sqrt{\left(\frac{n_x}{l_x}\right)^2 + \left(\frac{n_y}{l_y}\right)^2 + \left(\frac{n_z}{l_z}\right)^2} \qquad (2.26)$$

where,
x, y, and z are the three box dimensions,
l is the designated dimension of the box,
n takes on all possible integer values ($n = 0, 1, 2, 3,\ldots$).

If the lowest modal frequency found by setting $n = 0$ for all but the longest box dimension is above the band of use, there will be no standing waves inside the enclosure. In the more common case of a woofer being used above the first mode frequency, acoustic absorption can be added to damp these unwanted resonances. It should be understood that the addition of large amounts of damping material can have an adverse effect on the box/port tuning, so a balance must often be struck between control of standing waves and optimal low-frequency response alignment.

2.10.2.3 Measurement of Thiele-Small Parameters

The most accurate way of determining f_S is by observation of a Lissajous pattern on an oscilloscope of voltage versus current with the speaker in free air. When the pattern collapses to a straight diagonal line, the phase of the impedance is zero, which indicates resonance. Once this condition has been achieved, the frequency should be measured with a frequency counter that is accurate to 0.1 Hz or better.

V_{AS} may be determined with the loudspeaker suspended in free air as follows:

1. Find total moving mass (m) by attaching an extra mass (M_X) to the cone (as close to the voice coil as possible) and observing the new resonant frequency, f_{SX}. M_X can be putty or clay; measure m_X accurately. The total mass can then be found by the equation

$$m' = \frac{M_x}{\left(\dfrac{f_S}{f_{SX}}\right)^2 - 1} \qquad (2.27)$$

2. The suspension spring constant is then

$$k = \left(2\pi f_s\right)^2 m' \tag{2.28}$$

3. From the effective diaphragm area of the loudspeaker, S_D, V_{AS} is given by

$$V_{AS} = \frac{\rho_0 c^2 S_D{}^2}{k} \tag{2.29}$$

One means for determining the effective diaphragm area is to excite the woofer near its resonant frequency and observe its motion using a strobe light tuned almost, but not exactly, to the frequency of excitation. The resulting apparent slow motion of the woofer will allow an accurate determination of the portion of the cone that is moving.

Q_T may be found using the following procedure:

1. Determine the impedance of the woofer, Z_{max}, at its resonant frequency. This is simply the applied voltage divided by the current in the coil.
2. Identify the frequencies, f_+ and f_-, above and below f_s, respectively, at which the magnitude of the applied voltage divided by the magnitude of the current in the coil is equal to.
3. Mechanical Q is given by

$$Q_m = \left(\frac{f_0}{f_+ - f_-}\right)\sqrt{\frac{Z_{max}}{R_e}} \tag{2.30}$$

4. The total Q is then

$$Q_1 = Q_m \left(\frac{R_e}{Z_{max}}\right) \tag{2.31}$$

2.10.3 Horns

Although there are numerous mathematical treatments of horns in the texts on acoustics, they all suffer from a common set of inadequacies: the models developed in the literature account for energy transmission inside the horn, but there are no closed-form solutions to the problem of horn directivity—i.e., the behavior of a horn's radiation outside the boundaries of the horn walls, where listeners are located. Modern horn designers have been far less concerned with optimizing acoustic loading than with creating desirable directivity character-istics, and the designs have without exception been derived empirically rather than analytically.

In an exponential horn, the cross-sectional area is given by

$$S = S_0 e^{mx} \tag{2.32}$$

where,

S_0 is the cross-sectional area at the horn's throat, or entry,

m_x is called the flare constant.

The radiation impedance of an exponential horn, assumed to be infinitely long for our purposes, is

$$Z_r = \frac{\rho_0 c}{S_0}\left[\sqrt{1 - \frac{m^2 c^2}{4\omega^2}} + j\frac{mc}{2\omega}\right]. \qquad (2.33)$$

The first term in the brackets is the radiation resistance, and the second is the radiation reactance. Of interest is the frequency at which the value of expression inside the square root becomes zero

$$\omega_c = \frac{mc}{2} \qquad (2.34)$$

or

$$f_c = \frac{mc}{4\pi} \qquad (2.35)$$

This is known as the *horn cutoff frequency*. The above theory predicts that no sound will be transmitted through the horn below this frequency. Clearly, this is not the case with real horns, so the theory contains one or more assumptions that are not met in practice. Note also that the second term in the brackets, the radiation reactance, goes to zero in the high-frequency limit.

2.11 LOUDSPEAKER TESTING AND MEASUREMENT

As with most other devices that transmit or process a signal containing information, measurement techniques have been developed specifically for the testing and evaluation of loudspeakers. Before the early 1980s, accurate, comprehensive testing of loudspeakers generally required expensive anechoic chambers or large outdoor spaces. Since that time, the advent of computer-based time-windowed measurements has revolutionized the field of acoustic instrumentation, particularly as regards the testing of loudspeakers.

2.11.1 Linear Transfer Function

One objective in testing a loudspeaker is to determine the linear portion of its characteristic transfer function (or, equivalently, impulse response). The most common means for acquiring this data is a spectrum analyzer. A spectrum analyzer applies a signal with known spectral content to the input of a system and processes the signal that appears at the output of the device to acquire the system's transfer function.

2.11.1.1 Spectrum Analysis Concepts

All spectrum analysis techniques are subject to a set of general constraints imposed by the mathematical relationship between time and frequency. It is useful to have a feeling for these constraints when gathering or evaluating loudspeaker data. Time and frequency are the mathematical inverses of each other. A signal that has only one frequency must exist for all time and, conversely, a signal that exists for a finite amount of time must contain multiple frequencies. A signal that exists only within a known time interval—i.e., at all times before time t_0 the value of the signal is zero and at all times after time t_1 the value is zero—can only contain frequencies given by the expression:

$$f = N \frac{1}{(t_1 - t_0)} \tag{2.36}$$

where,
N is an integer.

The frequency corresponding to $N = 1$ gives the best (lowest) frequency resolution that is possible in a test conducted for that precise time interval. All other frequencies will be integer multiples of this frequency. In order to have infinitesimally small frequency resolution (i.e., perfectly resolved frequency data), a test would have to be conducted for an infinite amount of time. It follows that all realizable response tests have a limit on their frequency resolution.

The effect of frequency resolution on a transfer function measurement is to smooth the appearance of a plot of the results, thereby possibly obscuring some of the details of the transfer function. This smoothing is present to some degree in all transfer function measurements. In the case of electronic devices, transfer functions are typically well behaved enough that the frequency resolution of a response test does not cause meaningful loss of detail. With loudspeakers, the opposite is often true: a loudspeaker's transfer function often has so much fine structure that a practical test will noticeably smooth out the peaks and dips in the speaker's response. The degree to which this fine structure is audibly significant is a matter of some controversy. As a result, there is no widespread agreement in the industry on the minimum desirable frequency resolution in testing loudspeakers.

2.11.2 Chart Recorders

Prior to the advent of computer-based measurement systems, the most commonly employed loudspeaker measurement instrumentation comprised a strip chart recorder and a signal sweep generator. The two devices are synchronized such that, for a given frequency in the sweep, the pen on the recorder is in the appropriate x (frequency) position on preprinted graph paper. The pen's y position would correspond to the amplitude of the signal received from the test microphone, and therefore, hopefully, to the amplitude response of the speaker at that frequency.

If the y amplifier is logarithmic, then the amplitude will be expressed in decibels. As common as the strip-chart measurement technique was prior to the 1980s, it had several prominent disadvantages:

1. There is no means of measuring a loudspeaker's phase response with this technique.
2. The measurement is incapable of discriminating between direct sound from the device under test and sound that is reflected from surfaces in the test environment. This necessitated the construction of very costly anechoic chambers. Even in such a chamber, the inclusion of some reflected sound in a strip-chart type measurement is unavoidable.
3. The measurement technique does not isolate the linear portion of a loudspeaker's transfer function. Distortion products are simply added to the amplitude of the loudspeaker's transfer function at the fundamental frequency that excites them.
4. There is no direct, accurate way to determine or control the frequency resolution of a strip chart measurement. Reducing pen speed and/or increasing chart (and sweep) speed have the effect of reducing frequency resolution, or smoothing, the data, but the degree to which this has taken place is not always apparent.
5. Data from this form of measurement is only generated in hard copy format.
6. This measurement technique provides no ready means to compensate for propagation delay: the time required for sound to travel from the loudspeaker to the test microphone.
7. Since there is no means for distinguishing between the output signal from the loudspeaker and ambient noise, the test environment must be quiet.

2.11.3 Real Time Analyzers

Although initially developed to measure the response of sound systems in their operating environments, real time analysis has also been used to measure loudspeaker response in controlled environments. With this testing technique, a pink noise signal is applied to the loudspeaker. Pink noise is a random signal that contains equal energy for each unit of logarithmic frequency—e.g., for each octave or fraction thereof. The signal from the test microphone is applied to a series of bandpass filters of constant percent-octave bandwidth, each of which is tuned to a different band center, and the averaged output level of each filter is displayed, either on a CRT, LCD, or LED display. The display represents, within the limitations due to the measurement technique and the test environment, the amplitude response of the loudspeaker. Because the frequency content of a random signal has small fluctuations over time, the display may be averaged, or integrated, to produce a stable graph. Real time analysis suffers from the same general disadvantages as the chart recorder method of measurement.

2.11.4 Time-Windowed Measurements

The development of inexpensive computers has literally revolutionized the field of acoustic instrumentation. This is largely the result of the computer's ability to process and store large amounts of signal data. With a computer-based measurement system, processing and display of the data can be accomplished at any time after the raw data has been taken.

The effects of time windowing are present in any signal measurement, since the measurement must be initiated and completed in a finite amount of time (window). In digital measurement systems, however, the exact size of the time window, and therefore the resultant tradeoffs in resolution, are more directly controllable. There are two general approaches to acquiring a loudspeaker's transfer function via time-windowed measurements: measurement of the device's impulse response (time-domain measurement) or acquisition of the device's transfer function in the frequency domain.

2.11.4.1 Measurement of the Impulse Response

One form of input signal that is highly useful as an excitation for test and analysis purposes is an impulse. Mathematically, the signal is described by a Dirac delta function. Descriptively, an impulse is a voltage "spike" of very short duration and relatively large amplitude. An interesting property of an impulse is that it contains all frequencies at the same level. The impulse response of a loudspeaker, via the Fourier transform (or fast Fourier transform, FFT, as implemented in computer-based instruments), gives the speaker's transfer function. It is this equivalency via transform of the impulse response and transfer function that allows us to fully characterize a two-port device through measurement taken in only one domain or the other (time or frequency).

If a loudspeaker is excited with an impulse, the signal from a suitably well-behaved test microphone placed in front of the speaker will represent the loudspeaker's impulse response at that point. This signal may be digitized and postprocessed to yield the frequency-domain transfer function, as well as a number of other functions. The sampling process takes place over a fixed amount of time (the time window), and its initiation may be delayed to remove the effect of time required for sound to travel from the loudspeaker to the test microphone (propagation delay). Additionally, the length of the time window may be chosen so as to reject reflections from the room in which the measurement is being made. This is termed a *quasianechoic* measurement, and its availability has made it possible to acquire accurate direct-field response data on loudspeakers without an anechoic chamber.

The mathematics of Fourier series requires that the signal value be zero at the beginning and end of the time window. Since this condition is not generally satisfied, a window function is applied to the sampled data to force the endpoints of the window to zero. The effect of the window function is to create inaccuracies in the calculated transfer function, but these are generally much less than the spectral inaccuracies that would result from unwindowed (truncated) data. Various types of window functions may be used, including square (equivalent to unwindowed data), Gaussian, Hamming, and Hanning. Each has its advantages and drawbacks.

Among the disadvantages of impulse excitation is that of SNR. Because the impulse is of short duration, quiescent noise in the test environment can easily corrupt the test data. One means of reducing the effect of background noise is to average the results of multiple tests. If the noise is random, and therefore uncorrelated with the test signal, each doubling of the number of averages has the potential of reducing the relative noise level by 3 dB. Averaging increases the amount of time required to acquire the data.

Another disadvantage of impulse excitation is that it provides no means of identifying nonlinearities (distortion) in the loudspeaker. Distortion products appear no differently to the analyzer than the linear portion of the device's response.

2.11.4.2 Maximum Length Sequence Measurements

A variation on the method of impulse excitation is called maximum length sequence, or MLS, testing. In this form of measurement, the excitation signal is a series of pulses that repeats itself. The loudspeaker's impulse response is derived by calculating the cross-correlation between the input and output signals. This excitation signal has the advantage of producing higher average signal levels than an impulse, therefore improving the signal/noise performance of the test.

Additionally, the effects of certain types of distortion can be reduced substantially by running a series of tests employing a strategically chosen set of signal sequence lengths. To be accurate, an MLS test must be configured such that the duration of the signal sequence exceeds the length of the impulse response of the device under test. In the case of loudspeaker measurements, the impulse response of the acoustic environment must be accounted for in order to satisfy the requirement.

2.11.4.3 Other FFT-Based Measurements

Yet another variation on the FFT technique is a type of measurement that can use an arbitrary signal as the excitation signal. This form of measurement is a dual-channel FFT measurement, and the variations on this technique have several common elements. The technique involves sampling the signal at a point in the chain prior to the input of the loudspeaker (input), as well as the signal from a test microphone (output). The output may be sampled at a later time in order to account for the time required for sound to propagate from the loudspeaker to the test microphone. As with MLS testing, cross-correlation between input and output signals will yield the impulse response of the loudspeaker. It is also possible to perform an FFT on both input and output signals and obtain the transfer function of the loudspeaker by complex division.

The dual-channel approach has the advantage of allowing a wide range of signals, including music, to be employed as excitation. The commercially available implementations of this technique incorporate several refinements of the basic procedure described above, and these systems offer the possibility of measuring the response of a sound system while it is in operation.

SNR is a possible issue with this form of measurement, so averaging is generally performed to improve the accuracy of the results. Additionally, the spectrum of the input signal may not contain sufficient energy at all frequencies to sufficiently excite the system under test. For this reason, a coherence function is used to indicate those frequency ranges where the signal energy is insufficient to yield good results.

2.11.5 Swept Sine Measurements

Although the chart recorder is a swept sine measurement, it fails to take advantage of all the possibilities offered by the use of a sweep (also known as a *chirp*) as a test signal. Dick Heyser developed and patented a technique known as time delay spectrometry, or TDS. In TDS, the analyzer's receiving circuitry employs a bandpass filter, the center frequency of which is swept in synchronicity with the frequency of the signal applied to the loudspeaker. A delay may be applied to the sweep of the bandpass filter to account for the amount of time required for sound to propagate from the device under test to the microphone, hence the name of the technique.

The bandpass filter will reject frequencies that are displaced by some amount from its center frequency. If the receive delay for the filter is chosen appropriately, the analyzer will admit the direct signal from the loudspeaker, while simultaneously rejecting signals that have been reflected from environmental surfaces, thereby traveling a longer path and arriving later than the direct signal. The effect of this ability to reject unwanted reflections is the creation of a time window, even though the data is taken in the frequency domain. Additionally, the bandpass filter attenuates broadband noise by a much greater amount than it does the direct signal from the loudspeaker.

Due to the inherently high SNR of TDS, averaging of multiple tests is usually unnecessary. The number of samples analyzed is also not a function of the time window, as it is with an FFT-based analyzer. Furthermore, the bandpass filter removes distortion products, so TDS is intrinsically more capable of separating the linear transfer function from distortion products. It is also possible to use the technique to track specific harmonics while rejecting the fundamental.

Loudspeaker test instrumentation is more powerful and less expensive now than at any other time in the history of loudspeakers. While the current situation makes it possible to gather ever more detailed information about the behavior of loudspeakers, it is important to keep in mind the basics of instrumentation and spectrum analysis. This awareness will assist in identifying loudspeaker data that is suspect or incomplete.

REFERENCE

1. R. H. Small, "Constant-Voltage Crossover Network Design," *Journal of the Audio Engineering Society*, Vol. 19, no. 1, January 1971.

BIBLIOGRAPHY

A. A. Janszen, "An Electrostatic Loudspeaker Development," *J. Audio Eng. Soc.*, Vol. 3, no. 2, April 1955.

A. H. Benade, *Fundamentals of Musical Acoustics* Oxford, University Press, 1976.

A. N. Theile, "Loudspeakers in Vented Boxes: Part I & II," *J. Audio Eng. Soc.*, Vol. 19, no. 5, and 6, May and June 1971.

A. Wood, *Acoustics*, New York: Dover Publications, Inc., 1966.

Allan D. Pierce, *Acoustics: An Introduction to Its Physical Principles and Applications*, New York: McGraw-Hill, 1981.

B. N. Locanthi, "Application of Electric Circuit Analogies to Loudspeaker Design Problems," *J. Audio Eng. Soc.*, Vol. 19, no. 9, October 1971.

C. A. Henricksen, "Heat Transfer Mechanisms in Loudspeakers: Analysis Measurement and Design," *J. Audio Eng. Soc.*, Vol. 35, no. 10, October 1987.

C. A. Henricksen, "Phase Plug Modeling and Analysis: Radial versus Circumferential Types," presented at the 59th Convention of the AES, preprint 1328 (F-5).

C. R. Hanna and J. Slepian, "The Functions and Design of Horns for Loudspeakers," *J. Audio Eng. Soc.*, Vol. 25, no. 9, pp. 573–585, September 1977.

D. B. Keele Jr., "Low-Frequency Loudspeaker Assessment by Nearfield Sound-Pressure Measurements," *J. Audio Eng. Soc.*, Vol. 22, no. 3, pp. 154–162, April 1974.

D. B. Keele Jr., "What's So Sacred about Exponential Horns?" AES preprint 1038 (F-3).

D. B. Keele Jr., "Low-Frequency Horn Design Using Thiele/Small Driver Parameters," presented at the 57th convention of the AES, preprint no. 1250 (K-7), May 1977.

D. B. Keele Jr., "Effective Performance of Bessel Arrays," AES preprint 2846 (H1) presented at the 87th Convention of the AES, October 1989.

G. L. Augsburger, "Electrical versus Acoustical Parameters in the Design of Loudspeaker Crossover Networks," *Journal of the Audio Engineering Society*, Vol. 19, no. 6, June 1971.

H. F. Olson, *Acoustical Engineering* Princeton: Van Nostrand, 1976.

H. F. Olson, *Music, Physics and Engineering,* New York: Dover Publications.

H. Suzuki and J. Tichy, "Diffraction of Sound by a Convex or a Concave Dome in an Infinite Baffle," *Journal of the Acoustical Society of America*, preprint 0 (5), November 1981.

J. L. Bernstein, *Audio Systems*, New York: John Wiley & Sons, 1966.

J. M. Kates, "Radiation from a Dome," *J. Audio Eng. Soc.*, Vol. 24, no. 9, November 1976.

J. R. Ashley and A. L. Kaminsky, "Active and Passive Filters as Loudspeaker Crossover Networks," *J. Audio Eng. Soc.*, Vol. 19, no. 6, June 1971.

J. R. Ashley and M. D. Swam, "Experimental Determination of Low-Frequency Loudspeaker Parameters," *J. Audio Eng. Soc.*, Vol. 17, no. 5, October 1969.

J. R. Wright, "Fundamentals of Diffraction," *J. Audio Eng. Soc.*, May 1997.

John Borwick (Editor), *Loudspeaker and Headphone Handbook*, London: Butterworths, 1988.

John Vanderkooy, "A Simple Theory of Cabinet Edge Diffraction," *J. Audio Eng. Soc.*, December 1991.

K. O. Johnson, "Single-Ended Wide-Range Electrostatic Tweeters with High Efficiency and Improved Dynamic Range," *J. Audio Eng. Soc.*, July 1964.

L. E. Kinsler and A. R. Frey, *Fundamentals of Acoustics*, New York: John Wiley & Sons, 1962.

L. L. Beranek, *Acoustics*, New York: McGraw-Hill, 1954.

N. W. McLachlan, *Loudspeakers; Theory, Performances, Testing, and Design*, Corrected Edition, New York: Dover Publications, 1960.

P. M. Morse, *Vibration and Sound*, New York: McGraw-Hill, 1948.

P. W. Klipsch, "A Low-Frequency Horn of Small Dimensions," *Journal of the Acoustical Society of America*, October 1941.

P. W. Klipsch, "Modulation Distortion in Loudspeakers," *J. Audio Eng. Soc.*, Vol. 17, no. 2, April 1969. (See Parts II and III, *J. Audio Eng. Soc.*, Vol. 18, no. 1, February 1970 and Vol. 20, no. 10, December 1972.)

P. W. Klipsch, "Modulation Distortion in Loudspeakers Part II," *J. Audio Eng. Soc.*, Vol. 18, no. 1, February 1970.

R. F. Allison, "The Influence of Room Boundaries on Loudspeaker Power Output," *J. Audio Eng. Soc.*, Vol. 22, no. 5, June 1974.

R. G. Brown, *Lines Waves and Antennas*, New York: The Ronald Press Co., 1961.

R. H. Small, "Direct-Radiator Loudspeaker System Analysis," *J. Audio Eng. Soc.*, Vol. 20, no. 5, pp. 383–395, June 1972.

R. H. Small, "Closed Box Loudspeaker Systems; Part I: Analysis," *J. Audio Eng. Soc.*, Vol. 20, no. 10, pp. 798–808, December 1972.

R. H. Small, "Vented Box Loudspeaker Systems; Part I, II, III, and IV," *J. Audio Eng. Soc.*, Vol. 15, nos. 5, 6, 7, and 8; June, July, August, and September 1973.

R. H. Small, "Phase and Delay Distortion in Multiple-Driver Loudspeaker Systems," *J. Audio Eng. Soc.*, Vol. 19, no. 1, January 1971.

R. Heyser, "Loudspeaker Phase Characteristics and Time Delay Distortion Part 1," *J. Audio Eng. Soc.*, Vol. 17, no. 1, p. 30, January 1969.

R. Heyser, "Loudspeaker Phase Characteristics and Time Delay Distortion Part 2," *J. Audio Eng. Soc.*, Vol. 17, no. 2, p. 130, April 1969.

S. H. Linkwitz, "Active Crossover Networks for Non-Coincident Drivers," *J. Audio Eng. Soc.*, Vol. 24, no. 1, p. 2, January/February 1976.

S. Ishii and K. Takahashi, "Design of Linear Phase Multi-Way Loudspeaker System," AES 52nd Convention, October 1975, preprint 1059.

Loudspeaker Cluster Design

Ralph Heinz

3.1 WHY ARRAY?

For the purposes of this discussion we can define a loudspeaker array as *a group of two or more full-range loudspeaker systems, arranged so their enclosures are in contact.* System designers use arrays of multiple enclosures when a single enclosure cannot produce adequate sound pressure levels, when a single enclosure cannot cover the entire listening area, or both. These problems can also be dealt with by distributing single loudspeaker systems around the listening area, but most designers prefer to use arrays whenever possible because it is easier to maintain intelligibility using a sound source that approximates a point source than by using many widely separated sources.

3.2 ARRAY PROBLEMS AND PARTIAL SOLUTIONS: A CONDENSED HISTORY

First-generation portable sound systems designed for music used a very primitive form of array: they simply piled up lots of rectangular full range speaker systems together, with all sources aimed in the same direction, in order to produce the desired SPL. This type of array produced substantial interference, because each listener heard the output of several speakers, each at a different distance. The difference in arrival times produced peaks and nulls in the acoustic pressure wave at each location, and these reinforcements and cancellations varied in frequency depending on the distances involved. So although the system produced the desired SPL, the frequency response was very inconsistent across the coverage area. Even where adequate high-frequency energy was available, intelligibility was compromised by multiple arrivals at each listening location.

Second-generation systems incorporated compression drivers and horn-loading techniques derived from cinema sound reinforcement and used for large-scale speech-only systems (the original meaning of public address). When two or three of these horns were incorporated in a single enclosure with trapezoidal sides that splayed the horns away from each other, the first arrayable systems were introduced to the marketplace. These products promised to eliminate lobing and dead spots (peaks and nulls) and to drastically reduce comb filtering (interference). They did improve performance over the stack of rectangular enclosures loaded mainly with direct radiating cones. But frequency response across the coverage area remained inconsistent. In addition to the midrange and high-frequency variations across the coverage area of the array, low-frequency output varied from the front to the rear and side to side. Low-frequency energy was focused along the longitudinal axis of the array and close to it, producing a "power alley" that gave the seats with the best views the worst sound, Fig. 3.1.

3.3 CONVENTIONAL ARRAY SHORTCOMINGS

As we said in the first paragraph, the performance advantages of the array (whether horizontal or vertical) derive from its ability to approximate a perfect acoustical point source. But even the smallest arrays typically include three or more loudspeaker enclosures, each with two or three separate acoustic centers of its own. It's easy to appreciate that getting all those discrete sources to behave like a theoretical point source is difficult in practice. Signal processing solutions attempt to compensate for the difference between theory and reality by sacrificing the coherency of the electronic signal. They apply frequency shading and/or micro-delays to the signals sent to different enclosures, in order to ameliorate the acoustic problems. These approaches are costly, complicated, and often meet with limited success.

A rigorous analysis of the acoustical physics can point the way toward a practical, physical solution. First, consider what is probably the most common arrayable system in use today: $60° \times 40°$ horns in enclosures with $15°$ trapezoidal sides, Fig. 3.2.

FIGURE 3.1   A typical second-generation loudspeaker cluster. Even when a single enclosure is designed to resemble a point source, multiple enclosures will always interfere with each other when connected to a coherent audio signal.

Tight-packing three of these systems with their 15° sides touching produces a 30° splay between the horns, for a total included angle of 120°. At first glance, this seems like an ideal alignment. But the EASE interference predictions in Fig. 3.3 show the familiar and clearly audible problems with this configuration: significant interference above 1 kHz, with variations of 8 dB–9 dB depending on the angle. On axis, there is about 10 dB of gain at frequencies below 1 kHz.

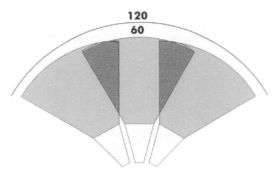

FIGURE 3.2   A very common array uses three 60° × 40° horns in enclosures with 15° trapezoidal sides; tight-packed, this array produces substantial overlap and interference between adjacent horns.

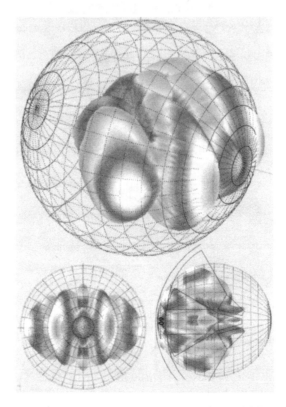

FIGURE 3.3   The interference patterns shown above were produced by tight-packing three array-able loudspeakers using 60° × 40° constant directivity horns in enclosures with 15° trapezoidal sides. While this is an improvement over a pile of direct radiating transducers, it is far from the ideal point source array.

Where maximum SPL is the main consideration, this type of array will deliver acceptable performance. When the front-of-house mix position can be located on the axis of left and right arrays, they can usually be tweaked to deliver acceptable reproduction in this limited area. Other areas of the house, including the high roller seats up front, will suffer.

The interference patterns displayed in Fig. 3.3 can be reduced by widening the splay between cabinets to 30°, as illustrated in Fig. 3.4. This array will not look as pretty as the first, but it does have much more even response across the coverage area, Fig. 3.5. At 2 kHz and 4 kHz, the individual horns are clearly discernible in the ALS-1 predictions. Also note that the seams between the horns become deeper with increasing frequency.

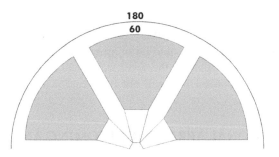

FIGURE 3.4   Widening the splay between horns reduces interference and widens the coverage angle to 180°, but reduces forward gain. As always, energy is conserved.

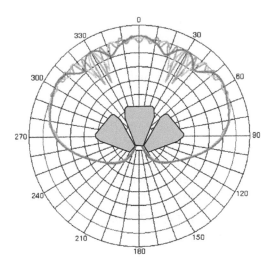

FIGURE 3.5   ALS-1 interference predictions for a wider splay show reduced interference, but the three horns are clearly apparent at higher frequencies.

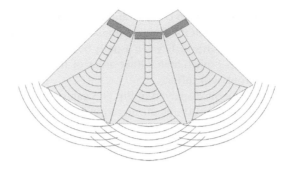

FIGURE 3.6   The acoustic pressure wave expands as a sphere, and multiple spherical sections will always overlap unless they originate from a common center.

Figure 3.6 shows why there will always be interference with conventional horn arrays (whether they are enclosed in arrayable cabinets with trapezoidal sides or mounted in free air). As the wave fronts radiate from points of origin that are separated in space, they will always create some interference at the coverage boundaries.

3.4 CONVENTIONAL ARRAY SHORTCOMING ANALYSIS

For an array in far field, dependence on angle is

$$SPL(\theta) = 10 \log P_0{}^2 \, \text{dB} \tag{3.1}$$

For a distance to the listening area very much larger than the array dimensions, let the sound pressure P be the real part of

$$P(\theta) = A_i(\theta)^{j(\omega\tau - kS_i)} \tag{3.2}$$

where,
P is the sound pressure,
ω is the angular frequency,
$A_i(\theta)$ is a function of the angle between the array longitudinal axis and the direction of the distant listening point. It gives the ratio of the sound pressure due to the source as a ratio of its on-axis value at the same distance.

For the ith source shown in Fig. 3.7, assuming identical sources, the pressure contribution is given by:

$$P_i = A_i(\theta)^{j(\omega\tau - kS_i)} \tag{3.3}$$

where,

k is $2\pi/\lambda = 2\,\lambda fc$,

λ is the wavelength,

f is the frequency,

c is the speed of sound,

S_i is the distance by which the path length from the ith source to the distant point exceeds the distance from the origin to that point.

For an array of n sources, the total pressure P is given by:

$$P(\theta) = \sum_{i=1}^{n} A_i q e^{j\omega\tau - kS_i}$$
$$= \varepsilon^{j\omega\tau} \sum_{i=1}^{n} A(\theta) \varepsilon^{j\omega\tau kS_i} \tag{3.4}$$

The square of the pressure amplitude is given by:

$$P_0^{\,2}(\theta) = \left[\sum_{i=1}^{n} A_i \theta kS_i\right]^2 + \left[A_i(q\theta)\sin(kS_i)\right]^2 \tag{3.5}$$

where,

$A_i(\theta)$ is $A_i(\theta - \alpha_i)$.

For a circular arc array, the additional path length S_i as shown in Fig. 3.7, for the ith source at radius R and angle α is given by:

$$-S_i(\theta) = R_i \cos(\theta - \alpha_i) \tag{3.6}$$

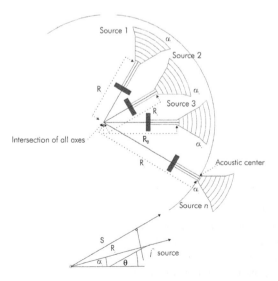

FIGURE 3.7   For a circular arc array, the additional path length S_i is as shown.

Therefore, the smaller R_i is, the smaller the S_i differences, and the less the interference between sources. Ideally, $R = 0$ for all sources. As R approaches 0, the interference will become less audible and frequency response across the array's intended coverage area will become more uniform.

3.5 COINCIDENT ACOUSTICAL CENTERS: A PRACTICAL APPROACH

Clearly, the ideal solution is to collocate all the acoustic points of origin, as shown in Fig. 3.8. We could achieve this by stacking the horns vertically, but this would solve the problem in the horizontal plane by creating a worse situation in the vertical (front to back) direction. Figure 3.9 shows a more realistic approximation that takes into account the physical constraints of loudspeaker design (the dimensions of the transducers, horns, enclosure walls, etc.). Because the acoustic sources are real physical objects, we cannot reduce R_i to 0. But we can get close enough to make measurable, audible improvements in the performance of the multi-enclosure array.

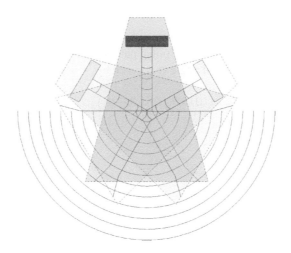

FIGURE 3.8   The acoustic ideal—colocating the acoustic centers of all horns is not a practical possibility.

3.5.1 TRAP Horns: A New Approach

Figure 3.9 implies that the way to minimize R_i—and the resultant interference—is to move the acoustic centers as far to the rear of the enclosure as possible. We can attempt to minimize the size of the drivers, for instance by using high-output magnetic materials such as neodymium. But the biggest obstacle to coincident acoustic centers is the horn itself. This is because typical constant

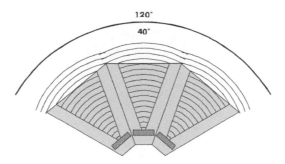

FIGURE 3.9 Because drivers and enclosures are physical objects, the acoustic centers of TRAP horns are not perfectly coincident but they are close enough to achieve measurable and audible reductions in interference.

directivity horns exhibit astigmatism: their apparent points of origin are different in the horizontal and vertical planes. In order to create a wider coverage pattern in the horizontal plane, the apparent apex is moved forward, while the vertical apex is farther to the rear because its coverage pattern is usually narrower. This is certainly the case with the most popular horn patterns in use today: 60° × 40° and 90° × 40°. One approach to a solution, then, is to rotate the horn and use the vertical apex of the horn in the horizontal plane. By doing so, we are effectively moving the acoustic center as far to the rear of the cabinet as possible. This technique when combined with cabinet design that minimizes the space between adjacent drivers in an array, while matching the trapezoidal sides with the opening angle of the horn, creates a system capable of minimal interference in the frequency range where the horn is effective. This forms the basis for what I call the True Array Principle by Renkus Heinz.

Subsequent refinements to the horn flare itself have been awarded U.S. Patent #5,750,943. This Arrayguide topology goes even farther in locating the apparent acoustic origin toward the rear of the enclosure. To repeat, moving the acoustic centers to the rear minimizes R, the distance between acoustic points of origin within the array, and the resulting interference between array elements.

Figure 3.10 shows the ALS-1 predictions for the first generation of TRAP horns. It is clear that interference has almost disappeared.

Figure 3.11 shows measured EASE data for a three-wide array of TRAP40 enclosures. Frequency response is consistent in both vertical and horizontal planes within ±4 dB. This is an out of the box array, using no frequency shading or micro-delay to improve performance. Measured results don't track the predictions 100% because the actual pattern of the horns varies somewhat with frequency: first-generation TRAP horns maintain nominal coverage ±10° from 1 kHz to 4 kHz.

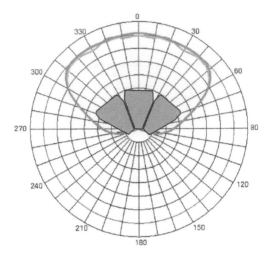

FIGURE 3.10   TRAP design produces truly arrayable systems with minimal destructive interference in the horns' pass band.

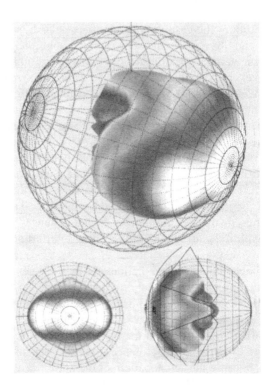

FIGURE 3.11   The TRAP array produces almost no measurable interference from a tight-packed three-wide cluster. This is because the three spherical wave-fronts produced by the three horns originate from a common acoustical center. Therefore they behave as a single acoustic unit, without overlap or interference.

3.5.2 TRAP Performance

Systems based on the True Array Principle can extend pattern bandwidth (the frequency range over which coverage varies less than ±5°) down to the frequency at which mutual coupling between adjacent cabinets ceases. TRAP systems are designed so that the enclosures provide optimum splay angles of 40° between the horns: the trapezoidal sides are therefore steeper than many other designs at 20° per side. The combination of symmetrical horns and steeper sidewall angles maintains coincident acoustic centers for all the elements in the array.

Note that moving the horizontal apex to the same location as the vertical results in a symmetrical 40° × 40° coverage pattern. This in turn requires the use of four enclosures to cover 160° with almost no variation in frequency response in the horizontal (side to side) plane. With 60° × 40° cabinets we could deliver sound to 180° of coverage, albeit with some quite audible variations.

There are other commercially available systems offering similar array performance to that described above. The ARC's system from French loudspeaker manufacturer L-Acoustic uses a type of path length equalizer to force the emerging wave front to conform to the opening angle of their horn and also puts the acoustic center behind the cabinet. In the case of ARC, the cabinet's trapezoidal side walls also serve as the waveguide for the high frequencies. As the waveguides opening angle matches that of the cabinet, this is certainly an elegant solution to creating minimum interference arrays at the frequency where the horn is effective.

In the KF900 series from EAW, simple phase horns for the mid and high frequencies put the acoustic center as close to the rear of the cabinet as possible, while their opening angles also match the trapezoidal sides of the enclosure. The relatively large size of the KF900 series enclosures and horns brings minimum interference performance to frequencies lower than those based on smaller waveguides. Remember, that this technique for minimum interference arrays, including the True Array Principle, only holds true for those frequencies where the horn is effective.

3.6 LOW-FREQUENCY ARRAYS: BENEFICIAL INTERFERENCE

In the preceding paragraphs, I outlined the parameters necessary to minimize destructive acoustic interference between adjacent cabinets or horns in an array. But these techniques are only beneficial at the frequencies where the horns are effective. Yet these very systems or horns are used at frequencies well below their directivity cutoff and lower, down to frequencies where the woofers piston size offers no directional control at all.

3.6.1 Horizontal Woofer Arrays: Maintaining Wide Dispersion

For our first example, let's look at the additional problems and opportunities we create when arraying small (12 inch woofer, 1 inch compression driver) full range loudspeaker enclosures as in Fig. 3.12. For a full range

FIGURE 3.12   TRAP arrays can be quite small; however, the size of the horns will determine the lower frequency limit at which the True Array Principle ceases to operate.

array module, there are three frequency zones that exhibit different wavelength related behavior. At the lowest frequencies, or longest wavelengths, these modules exhibit only beneficial interference or mutual coupling. Each additional module creates additional on-axis acoustic output. The opportunity here, is that less equalization is required to make the array's frequency response flat down to these lower frequencies as compared to a single cabinet.

A potential problem is created when the array becomes too wide, however. Four or five element arrays are wide enough as to become quite directional in the forward plane at those lower frequencies (20 Hz to roughly 500 Hz or more, dependent on the module). Without a signal processing scheme, this array cannot be equalized to have the same frequency response throughout its intended coverage. It will sound boomy in the middle and thin at its coverage extremes. A solution is to taper the length of the array in the horizontal plane in order to maximize horizontal dispersion of the lower frequencies. The entire array can be used for the lowest frequencies as the wavelengths are longest (20 Hz up to about 200 Hz), but at higher frequencies, as the wavelengths get shorter, the array length must also get shorter to maintain wide dispersion. This is achieved by low passing the outermost woofers of the array, such that only two or three max woofers are used at frequencies higher than this.

The second frequency zone that can be problematic in arrays based on full range modules, occurs at wavelengths where cabinet spacing no longer supports mutual coupling, and the horn has yet to attain its directivity cutoff. This typically applies to a small half octave range where adjacent cabinet spacing approaches a wavelength. Here we observe combinations of destructive and constructive interference at various observation points around the array intended coverage, causing frequency response variations greater than ±6 dB. Fortunately there is a signal processing technique that can minimize this effect. By simply notching this frequency range from every other cabinet with a cut equal to the greatest amount of variance (typically 6 dB of attenuation), and width equal to the bandwidth of the aberrations (typically half an octave), the frequency response variations throughout the arrays coverage can be minimized.

The third frequency zone of wavelength related behavior for arrays based on full range modules, is then at the frequencies above which the horn is effective. Let us assume that the horns depicted in Fig. 3.12 place the acoustic center towards the rear of the cabinets, and that their opening angle also matches that of the trapezoidal sides of the cabinet. Based on these assumptions, the array performance will exhibit minimum interference for frequencies above 1–2 kHz which happens to be the effective directivity cutoff of the horn. Each additional module simply adds additional coverage to the array.

3.6.2 Vertical Woofer Arrays

3.6.2.1 Directivity at Frequencies Where Size Makes Horns Impractical

Beneficial destructive interference sounds like an oxymoron, but there are several commercially available woofer arrays that take advantage of this very technique. By applying the fundamental physics described by Harry Olson, directional woofer arrays are now available that outperform large woofer horns.

When two point sources are superimposed on one another, their outputs simply add up in all directions. As the two point sources are spread apart, the output diminishes along the plane of separation due to phase cancellation. At exactly ½ wavelength, a pure null occurs, and we achieve the classic figure 8, dipole polar pattern. The current commercially available systems take advantage of this phenomena, directivity through off-axis attenuation, by placing woofers in a vertical array and spacing them to create this dipolar pattern at frequencies below which horns become too large.

Figure 3.13 is an example of one such array. Termed *Tri-Polar* by its designer Vance Breshears, it uses the vertical spacing between the three woofers with appropriate signal processing to maintain consistent low-frequency pattern control from 400 Hz down to below 100 Hz. One of the first systems available was developed by Craig Janssen, termed *Tuned Dipolar*, which uses two separate arrays. With drivers, spacing, and signal processing appropriate for their respective passbands Tuned Dipolar offers exceptional low-frequency pattern control over an extended bandwidth. Even subwoofers are now benefitting

FIGURE 3.13   Reference Point Array using four 40° × 40° mid-high enclosures and six low-frequency modules in Tri-Polar configuration for vertical pattern control, along with appropriate small full range systems for downfill.

from this type of technology. Meyer Sound is achieving cardioid patterns at lowest frequencies from its PSW-6, providing significant attenuation of those frequencies directly behind the enclosures.

3.7 LINE ARRAYS AND DIGITALLY STEERABLE LOUDSPEAKER COLUMN ARRAYS

For the communication between a source and a listener to be effective, it is important that the listener receive and comprehend the message. In large spaces where people gather, including auditoria, houses of worship, sports venues, transit terminals, and classrooms, often the acoustic requirements that enable effective speech are in conflict with the architectural needs of the spaces. When the acoustics of a venue cannot be altered to enable effective speech communication, designing a sound reinforcement system to do so can be a challenge.

Recent advances in efficient amplification and digital signal processing have enabled a new class of loudspeaker: the *digitally steerable column* or *line array* as it's often called. The acoustical and architectural benefits of these loudspeakers for sound reinforcement in highly reverberant or reflective environments will be shown.

We will discuss effective communications and define intelligibility and how to measure it both subjectively and objectively. We will look at architecture and acoustics and at reverberation and its effect on intelligibility in large public spaces. Finally we'll look at digitally steerable column arrays, their design considerations, and their performance and benefits when used in large reverberant spaces.

Some of the basic principles involved in voice communications are:

- In voice communications intelligibility is the capability of being understood.
- It assumes the existence of a communication process between a talker and a listener, or between a source and a listener.
- For the conveyance of meaning, the English language is highly dependent upon the effective receipt and comprehension of consonants. This is how we differentiate words based on similar vowels. For example, Zoo, Two, New.
- In terms of frequency response, speech ranges between 100 Hz and 8 kHz, with maximum energy around 250 Hz.
- In speech, the frequency range that conveys the most consonant information is the octave around 2 kHz.

3.7.1 What Affects Intelligibility

Major influences that affect intelligibility are:

- Elocution and pronunciation of the talker. It's hard to understand someone who mumbles under any condition.
- Hearing acuity of listener. An often overlooked influence, those with a hearing loss have trouble understanding what's being said.
- SNR. We've all been places where it was so noisy we couldn't understand what was being said.
- Direct to reverberant ratio. The higher the reverberation level, the more difficult it is to understand what's being said.
- Directivity of the loudspeaker or loudspeakers. Highly directional loudspeakers direct more of the sound onto the audience and less onto the reflective walls and ceilings.
- The number of loudspeakers. Larger numbers of loudspeakers translate into more acoustic energy being transmitted into the room and higher reverberation levels.

- Reverberation time. The longer the reverberation time, the more likely it will interfere with intelligibility.
- Distance of source to listener. The closer the listener is to the loudspeaker, the less likely reverberation will interfere.

Secondary influences are:

- Gender of talker.
- Microphone technique.
- Vocabulary and context of speech information.
- Direction of main sound to listener and/or direction of reflections and echoes.
- System fidelity, equalization, and distortion.
- Uniformity of coverage.

3.7.2 Measuring Intelligibility

3.7.2.1 Subjectively

Statistical tests with trained talkers and listeners can be the most reliable metric for determining the intelligibility of a system. To ensure that all speech sounds are represented in a test, Phonemically Balanced (PB) word lists are commonly used. These word lists can be as long as 1000 words. Tests using nonsense syllables or logatoms, and Modified Rhyme Tests are also used. These tests are very time consuming and are difficult to set up.

3.7.2.2 Objectively

Articulation Index. Articulation Index or AI was one of the first attempts to quantify intelligibility with measurements. AI is primarily concerned with the affect of noise on speech. The index ranges from 0 to 1 with 0 representing no intelligibility.

%ALcons. %ALcons or the articulation loss of consonants was developed by Peutz in Holland during the 1970's. %ALcons takes both noise and reverberation into account and is based on the importance of the octave around 2000 Hz in conveying consonant information. %ALcons uses a scale running downwards from 0 where 0 is perfect intelligibility, or 0% articulation loss.

Although Peutz used 2000 Hz as the center frequency and 2000 Hz is still the European standard, many acousticians in the USA prefer using 1000 Hz. As a general rule, %ALcons calculated at 1000 Hz show a higher articulation loss than ones calculated at 2000 Hz.

STI. STI or Speech Transmission Index considers the source/room/listener as a transmission channel and measures the reduction in modulation depth of a specialized test signal which replicates the burst nature of real speech. The STI scale ranges from 0 to 1, where 1 represents perfect intelligibility. STI is considered the most accurate of the intelligibility measures.

Evaluation	STI	%ALcons
Bad	0.20 to 0.34	24.3 to 57
Poor	0.35 to 0.50	11.3 to 24.2
Fair	0.51 to 0.64	5.1 to 11.2
Good	0.65 to 0.86	1.6 to 5.0
Excellent	0.87 to 1.00	0.0 to 1.5

Copied from The Audio System Designer Technical Reference by Peter Mapp and published by Klark Teknik.

3.7.3 Architecture and Room Acoustics

3.7.3.1 Reverberation

Reverberation is the persistence of sound in a space after the original sound has been removed.

RT60 is the measure for reverberation, and it is defined as the amount of time required for the average sound energy density in a space to decrease from its original value by 60 dB after the original sound has stopped.

The Sabine equation relates RT60 to the volume of a room with its surface area and the absorption coefficients of the materials applied to the surfaces.

As room volume increases relative to surface area and absorption coefficients, the RT60 increases.

As surface area and absorption increase relative to room volume, RT60 decreases. It is this persistence of sound that interferes with our comprehension of consonants and contributes towards degrading intelligibility. Table 3.1 shows the effect of reverberation on intelligibility.

TABLE 3.1 Intelligibility Comparison Chart

RT60	<1 s	Excellent intelligibility can be achieved.
RT60	1 to 1.2 s	Excellent to good intelligibility is possible.
RT60	1.2 to 1.5 s	Good intelligibility can be achieved.
RT60	>1.5 s	Careful system design is required.
RT60	>1.7 s	Limit for good intelligibility in large spaces.
RT60	>2 s	Very directional loudspeakers are required, intelligibility can have limitations.
RT60	>2.5 s	Intelligibility will probably have limitations.
RT60	>4 s	Highly directional loudspeakers will be required to achieve acceptable intelligibility.

3.7.4 Line Arrays

Figures 3.14 to 3.16 show the direct sound coverage of various loudspeakers in a sanctuary 100 ft × 65 ft. The chancel adds 20 ft to its length. The roof peaks at 52 ft. The room volume is roughly 250,000 ft^3. The room has plaster walls, wood ceiling, terrazzo floors, and empty wooden pews. This produces a RT60 of about 3.5 s.

Notice the high SPL levels on the walls and ceiling in the flown-horn array simulation. The high-frequency beaming of the mechanically tilted column

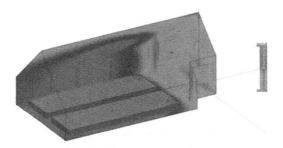

FIGURE 3.14   Flown large format horn array.

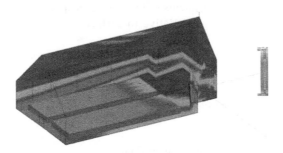

FIGURE 3.15   Mechanically tilted 4 meter column array.

FIGURE 3.16   Digitally steered column array.

array prevents good coverage of the front of the audience area. The digitally steerable column array covers only the audience area and has very little coverage on the walls and no coverage of the ceiling. Only the steered column array has acceptable (good to fair) intelligibility throughout the audience areas. Digitally steerable column arrays can offer superior coverage and they can provide improved D/R. They can provide improved intelligibility in highly reverberant spaces, plus they blend better with their surrounding architecture and are nearly invisible in use.

3.7.4.1 Digitally Steered Column Arrays

When the room size and volume are fixed and adding absorption to reduce the RT times is not an option, digitally steerable column arrays offer a new solution:

- They have the ability to be much more directional than the largest horns.
- The idea is not new; the concepts for these column arrays were described by Harry Olson in 1957. Only the implementation is new.
- The hardware required to implement these ideas is now available.
- Digital Signal Processing required is now a mature technology, very powerful and relatively inexpensive.
- Compact, highly efficient Class D amplifiers are capable of high-fidelity performance.

Line Arrays are not a new idea. Harry F. Olson did the math and described the directional characteristics of a continuous line source in his classic *Acoustical Engineering*, first published in 1940. Traditional column loudspeakers have always made use of line source directivity.

Simple line arrays (column arrays) are basically a number of drivers stacked closely together in a line, Fig. 3.17. Simple line arrays become increasingly directional in the vertical plane as the frequency increases. The spacing between drivers controls the high-frequency limits. The height (length) of the line array determines the low-frequency control limit. Figure 3.18 shows the line source directivity as described by Harry Olson in 1957.

The directivity of a line array is a function of the line length and the wavelength. As the wavelength approaches the line length, the array becomes omnidirectional, Fig. 3.19. Figure 3.20 shows the vertical dispersion pattern of a typical line array

3.7.4.2 Controlling High-Frequency Beaming

Simple line arrays become increasingly directional as the frequency increases, in fact, at higher frequencies they become too directional. The vertical directivity can be made more consistent by making the array shorter as the frequency increases by using fewer drivers. One amplifier channel and one DSP channel per driver make this possible.

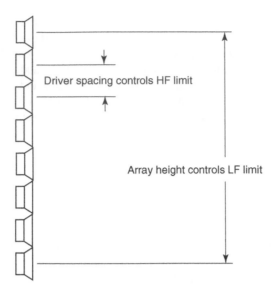

Driver spacing controls HF limit

Array height controls LF limit

FIGURE 3.17   Basic line array theory.

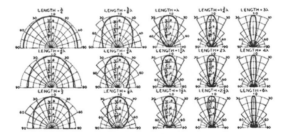

FIGURE 3.18   Directional characteristics of a line source as a function of the length and the wavelength. The polar graph depicts the sound pressure at a large fixed distance, as a function of angle. The sound pressure for the angle 0° is arbitrarily chosen as unity. The direction corresponding to the angle 0° is perpendicular to the line. The directional characteristics in 3D are surfaces of revolution about the line as an axis. (From Acoustical Engineering by Harry Olson.)

3.7.4.3 Beam Steering

The beam can be steered up or down by delaying the signal to adjacent drivers. DSP control also allows us to develop multiple beams from a single line array and individually steer these beams.

DSP control also allows us to move each beam's acoustic center up and down the column allowing us to create multiple beams and also steer the beam, Figs. 3.21 and 3.22.

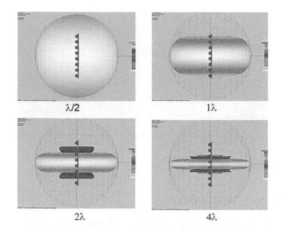

FIGURE 3.19   Simple line source directivity as a function of line length versus frequency.

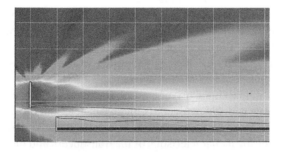

FIGURE 3.20   Typical line array vertical dispersion display.

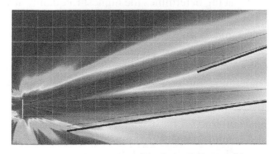

FIGURE 3.21   Vertical dispersion display showing multiple beam capability.

FIGURE 3.22   Graphic illustration of beam steering.

3.7.5 DSP-Driven Vertical Arrays

3.7.5.1 Acoustical, Electronic, and Mechanical Considerations

Practical examples are taken from the new Renkus Heinz IC Series Iconyx steerable column arrays. Iconyx is a steerable column array that combines very high directivity with accurate reproduction of source material in a compact and architecturally pleasing package, Fig. 3.23.

Like every loudspeaker system, Iconyx is designed to meet the challenges of a specific range of applications. Many of the critical design parameters are, of course, determined by the nature of these target applications. To understand the decisions that have been made during the design process we must start with the particular problems posed by the intended applications.

The function of individual driver control and DSP is to make more effective use of this phenomenon. No amount of silicon can get around the laws of acoustical physics. The acoustical properties of first-generation column loudspeakers are set by the acoustical characteristics of the transducers and the physical characteristics of the package:

1. The height of the column determines the lowest frequency at which it exerts any control over the vertical dispersion.
2. The inter-driver spacing determines the highest frequency at which the array acts as a line source rather than a collection of separate sources.
3. Horizontal dispersion is fixed and is typically set when the drivers are selected, because column loudspeakers do not have waveguides.
4. Other driver characteristics such as bandwidth, power handling, and sensitivity will determine the equivalent performance characteristics of the system.

One unfortunate corollary of these characteristics is that the power response of a conventional column loudspeaker is not smooth. It will deliver much more low-frequency energy into the room and this energy will tend to have a wider vertical dispersion. This can make the critical distance even shorter because the reverberant field contains more low-frequency energy, making it harder

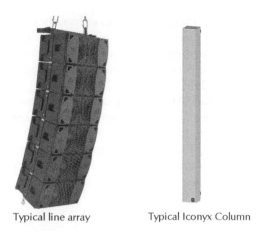

Typical line array   Typical Iconyx Column

FIGURE 3.23   Typical line array and a typical Iconyx Column.

for the listener to recognize higher-frequency sounds such as consonants or instrumental attack transients.

3.7.5.2 Point Source Interactions

3.7.5.2.1 Doublet Source Directivity

Doublet source cancels each other's output directly above and below, because they are spaced ½ wavelength apart in the vertical plane. In the horizontal plane, both sources sum. The overall output looks like Fig. 3.24.

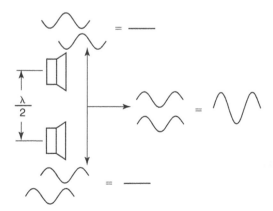

FIGURE 3.24   Output of a signal whose wavelength is ½ of the space between the two loudspeakers.

When two sources are ¼ wavelength apart or less, they behave almost like a single source. There is very slight narrowing in the vertical plane, Fig. 3.25a.

There is significant narrowing in the vertical plane at ½ wavelength spacing, because the waveforms cancel each other in the vertical plane, where they are 180° out of phase, Fig. 3.25b.

At one wavelength spacing the two sources reinforce each other in both the vertical and horizontal directions. This creates two lobes, one vertical and the other horizontal, Fig. 3.25c.

As the ratio of wavelength to inter-driver spacing increases, so do the number of lobes. With fixed drivers as used in line arrays, the ratio increases as frequency increases ($\lambda = c/f$ where f is the frequency and c is the speed of sound), Fig. 3.25d.

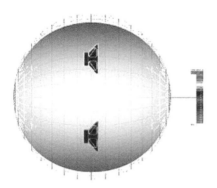

FIGURE 3.25a $\lambda/4$ (¼ wavelength).

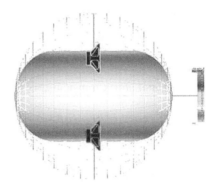

FIGURE 3.25b $\lambda/2$ (½ wavelength).

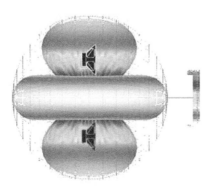

FIGURE 3.25c λ (1 wavelength).

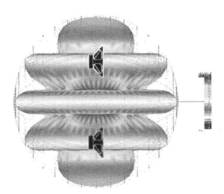

FIGURE 3.25d   Increased wavelength to inter-driver spacing.

3.7.5.2.2 Array Height versus Wavelength (λ)

Driver-to-driver spacing sets the highest frequency at which the array operates as a line source. The total height of the array sets the lowest frequency at which it has any vertical directivity.

Figures 3.26a through 3.26d show the effect of array height versus wavelength.

At wavelengths of twice the array height, there is no pattern control, the output is that of a single source with very high power handling, Fig. 3.26a.

As the frequency rises, wavelength approaches the height of the line. At this point there is substantial control in the vertical plane, Fig. 3.26b.

At higher frequencies the vertical beam width continues to narrow. Some side lobes appear but the energy radiated in this direction is not significant compared to the front and back lobes, Fig. 3.26c.

With still further vertical narrowing, side lobes become more complex and somewhat greater in energy, Fig. 3.26d.

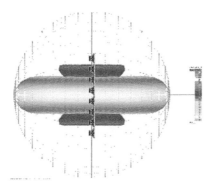

FIGURE 3.26a   Wavelength is twice the loudspeaker height.

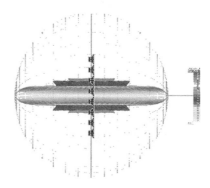

FIGURE 3.26b   Wavelength is the loudspeaker height.

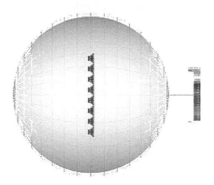

FIGURE 3.26c   Wavelength is one-half the loudspeaker height.

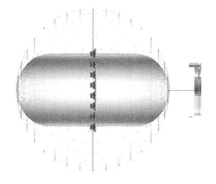

FIGURE 3.26d   Wavelength is one-fourth the loudspeaker height.

3.7.5.2.3 Inter-Driver Spacing versus Wavelength (λ)

The distinction between side lobes and grating lobes should to be maintained. Side lobes are adjacent to and radiate in the same direction as the primary lobe. Grating lobes are the strong summations tangential to the primary lobe. Side lobes will be present in any realizable line array, grating lobes form when the inter-driver spacing becomes less than ½ wavelength. It might also be good to point out that all of the graphics for this section are done using theoretical point sources.

Figures 3.27a through 3.27d show the effect of inter-driver spacing versus wavelength.

When the drivers are spaced no more than ½ wavelength apart, the array produces a tightly directional beam with minimal side lobes, Fig. 3.27a.

As the frequency rises, wavelength approaches the spacing between drivers. At this point, grating lobes become significant in the measurement. They may not be a problem, if most or all of the audience is located outside these vertical lobes, Fig. 3.27b.

At still higher frequencies, lobes multiply and it becomes harder to isolate the audience from the lobes or their reflections, Fig. 3.27c.

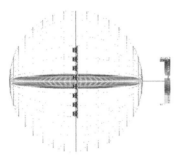

FIGURE 3.27a   Interspacing is one-half the wavelength.

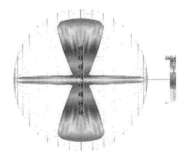

FIGURE 3.27b   Interspacing is one times the wavelength.

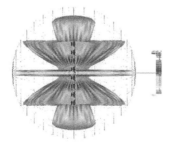

FIGURE 3.27c   Interspacing is two times the wavelength.

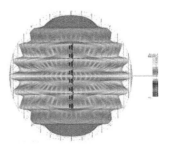

FIGURE 3.27d   Interspacing is four times the wavelength.

As inter-driver spacing approaches four times the wavelength, the array is generating so many grating lobes of such significant energy that its output closely approximates a single point source, Fig. 3.27d. We have come full circle to where the array's radiated energy is about the same as it was when array height was ½ λ. As shown in Fig. 3.26d, this is the high-frequency limit of line array directivity.

As real drivers are considerably more directional than point sources at the frequencies where grating lobes are generated, the grating lobes are much lower in level than the primary lobe, Figs. 3.28 and 3.29.

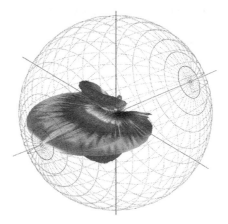

FIGURE 3.28   3D view of a second-generation Iconyx array at 4000 Hz.

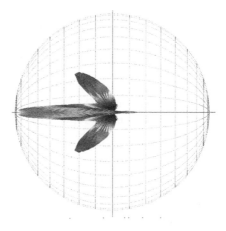

FIGURE 3.29   Side view of a second-generation Iconyx array at 4000 Hz.

3.7.6 Multichannel DSP Can Control Array Height

The upper limit of a vertical array's pattern control is always set by the inter-driver spacing. The design challenge is to minimize this dimension while optimizing frequency response and maximum output and do it without imposing excessive cost. Line arrays become increasingly directional as frequency increases, in fact, at high frequencies they are too directional to be acoustically useful. However, if we have individual DSP available for each driver, we can use it to make the array acoustically shorter as frequency increases—this will keep the vertical directivity more consistent. The technique is conceptually simple—use low-pass filters to attenuate drive level to the transducers at the top and bottom of the array, with steeper filter slopes on the extreme ends and more gradual slopes as we progress to the center. As basic as this technique is, it is practically impossible without devoting one amplifier channel and one DSP channel to each driver in the array.

A simplified schematic shows how multichannel DSP can shorten the array as frequency increases. For clarity, only half the processing channels are shown and delays are not diagrammed, Fig. 3.30.

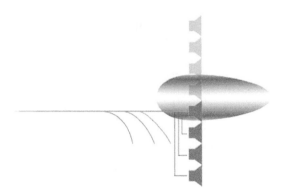

FIGURE 3.30   Multichannel DSP shortens the loudspeaker length.

3.7.7 Steerable Arrays May Look Like Columns But They Are Not

Simple column loudspeakers provide vertical directivity, but the height of the beam changes with frequency. The overall Q of these loudspeakers is therefore lower than required. Many early designs used small-cone full range transducers, and the poor high-frequency response of these drivers certainly did nothing to enhance their reputation.

3.7.7.1 Beam-Steering: Further Proof that Everything Old is New Again

As Don Davis famously quotes Vern Knudsen, "The ancients keep stealing our ideas." Here is another illustration from Harry F. Olson's *Acoustical Engineering*. This one shows how digital delay, applied to a line of individual sound

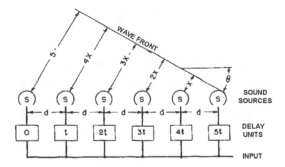

FIGURE 3.31   A delay system for tilting the directional characteristic of a line sound source. (From Acoustical Engineering by Harry Olson.)

sources, can produce the same effect as tilting the line source. It would be long after 1957 before the cost of this relatively straightforward system became low enough for commercially viable solutions to come to market, Fig. 3.31.

3.7.7.2 DSP-Driven Arrays Solve Both Acoustical and Architectural Problems

3.7.7.2.1 Variable Q

DSP-driven line arrays have variable Q because we can use controlled interference to change the opening angle of the vertical beam. The Renkus Heinz IC Series can produce 5°, 10°, 15°, or 20° opening angles if the array is sufficiently tall (an IC24 is the minimum required for a 5° vertical beam). This vertically narrow beam minimizes excitation of the reverberant field because very little energy is reflected off the ceiling and floor.

3.7.7.2.2 Consistent Q with Frequency

By controlling each driver individually with DSP and independent amplifier channels, we can use signal processing to keep directivity constant over a wide operating band. This not only minimizes the reverberant energy in the room, but delivers constant power response. The combination of variable Q, which is much higher than that of an unprocessed vertical array, with consistent Q over a relatively wide operating band, is the reason that DSP-driven Iconyx arrays give acoustical results that are so much more useful.

3.7.7.2.3 Ability to Steer the Acoustic Beam Independently of the Enclosure Mounting Angle

Although beam-steering is relatively trivial from a signal-processing point of view, it is important for the architectural component of the solution. A column mounted flush to the wall can be made nearly invisible, but a down-tilted column is an intrusion on the architectural design. Any DSP-driven array can be steered.

Iconyx also has the ability to change the acoustic center of the array in the vertical plane which can be very useful at times.

3.7.7.2.4 Design Criteria: Meeting Application Challenges

The previous figures make it clear that any line source, even with very sophisticated DSP, can control only a limited range of frequencies. However, by using full range coaxial drivers as the line source elements could make the overall sound of the system more accurate and natural without seriously compromising the benefits of beam-shaping and steering. In typical program material, most of the energy is within the range of controllable frequencies. Earlier designs radiate only slightly above and below the frequencies that are controllable. Thus much of the program source is sacrificed, without a significant increase in intelligibility.

To maximize the effectiveness of a digitally controlled line source, it's not enough to start with high-quality transducers. The Renkus Heinz Iconyx loudspeaker system uses a compact multichannel amplifier with integral DSP capability. The D2 audio module has the required output, full DSP control, and the added advantage of a purely digital signal path option. When PCM data is delivered to the channel via an AES/EBU or CobraNet input, the D2 audio processor/amplifier converts it directly into PWM data that can drive the output stage.

3.7.7.2.5 Horizontal Directivity is Determined by the Array Elements

Vertical arrays, including Iconyx, can be steered only in the vertical plane. Horizontal coverage is fixed and is determined by the choice of array elements. The transducers used in Iconyx modules have a horizontal dispersion that is consistent over a wide operating band, varying between 140° and 150° from 100 Hz to 16 kHz.

3.7.7.2.6 Steering is Simple—Just Progressively Delay Drivers

If we tilt an array, we move the drivers in time as well as in space. Consider a line array of drivers that is hinged at the top and tilted downward. Tilting moves the bottom drivers further away from the listener in time as well as in space. We can produce the same acoustical effect by applying progressively longer delays to each driver as we move from top to bottom of the array.

Again, steering is not a new idea. It is different from mechanical aiming—front and rear lobes steer the same direction.

3.7.7.2.7 BeamWare: The Software That Controls Iconyx Linear Array Systems

A series of low-pass filters can maintain constant beam width over the widest possible frequency range. The ideas are simple, but for the most basic Iconyx array, the IC16, we must calculate and apply 16 sets of FIR filters, and 16 separate delay times. If we intend to take advantage of constant inter-driver spacing to move the acoustical center of the main lobe above or below the physical center of the array, we must calculate and apply a different set of filters and delays. Theoretical models are necessary, but the behavior of real transducers is more complex than the model. Each of the complex calculations underlying the Iconyx beam-shaping filters were simulated, then verified by measuring actual arrays in our robotic test and measurement facility. Fortunately, the current generation of laptop and desktop CPUs are up to the task. BeamWare takes user input in graphic form (side section of the audience area, location and mounting angle of the physical array) and provides both a simulation of the array output that can be imported into EASE v4.0 or higher, and a set of FIR filters that can be downloaded to the Iconyx system via RS422 serial control. The result is a graphical user interface that delivers precise, predictable, and repeatable results in real-world acoustical environments.

Index

Printed and bound by CPI Group (UK) Ltd, Croydon, CR0 4YY

21/10/2024

01777097-0004